TECHNOCRACY
AT WORK

TECHNOCRACY
AT WORK

BEVERLY H. BURRIS

STATE UNIVERSITY OF NEW YORK PRESS

Published by
State University of New York Press, Albany

© 1993 State University of New York

For information, address State University of New York
Press, State University Plaza, Albany, N.Y., 12246

Production by E. Moore
Marketing by Fran Keneston

Library of Congress Cataloging-in-Publication Data

Burris, Beverly H., 1950-
    Technocracy at work / Beverly H. Burris.
        p.   cm. — (SUNY series the new inequalities)
    Includes bibliographical references and index.
    ISBN 0-7914-1495-7 (acid-free paper). — ISBN 0-7914-1496-5 (pbk.
    acid-free paper)
        1. Technocracy. I. Title. II. Series.
    HB87.B847 1993
    338'.06—dc20   92-24052
                                                                          CIP

10 9 8 7 6 5 4 3 2 1

This book is dedicated to Julia, who will be eleven in the year 2000, and to her generation, who will confront the problems of the twenty-first century. May this book play a role in defining the problems for them; the solutions will have to come from themselves.

CONTENTS

ACKNOWLEDGMENTS

This book represents the culmination of many years of pondering the relationship between knowledge and power, the effects of computerization, and the role of experts in advanced industrial societies. Along the way, many people have contributed to my thinking and have helped bring the book to fruition. Some of the following people read parts of the manuscript and made suggestions for revision, others made bibliographic suggestions, and others provided valuable research assistance: Mats Alvesson, Belinda Archuleta, Heidi Ballard, Greg Bischak, Jesse Dillard, Wolf Heydebrand, Randy Hodson, Jackie Hood, Michael Kennedy, Miguel Korzeniewicz, Louise Lamphere, Tom Mouck, Helen Muller, Diana Robin, David Stark, Bob Stern, Judith Thompson. It should be noted that some of these people made suggestions for revision that, for one reason or another, were not acted upon; therefore, none of them should be held responsible for existing deficiencies of this work.

A Faculty Scholars Award Sabbatical from the University of New Mexico enabled me to be released from teaching in 1988 and to begin the process of researching the book.

My daughter Julia was born when this book was half written. Her existence has delayed the completion of the book but, I believe, has made it a better book in the end. I know that she has made me a better person.

Special thanks are due to my husband, Rick. He did more than his share of the child care and housework so that I could write, and he put up with my being distracted and harried. More importantly, he believed in the importance of my finishing the book. If more men were like Rick, I know that more women would write books.

Chapter 1

INTRODUCTION

*Our faulty representations of some immense communicational
and computer network are themselves but a distorted figura-
tion of something even deeper, namely the whole world system
of present-day multinational capitalism.*
—*Frederick Jameson,* Postmodernism

The post-World War II period has witnessed a fundamental
restructuring of work organizations and labor markets around the
world. Empirical social science has described and documented these
trends and has begun to understand some of the causal factors implicated in the changes: the internationalization of the division of labor,
implementation of computerized technology and advanced information systems, intensified worldwide competition among firms and a
corresponding emphasis on innovation, expanded need (and capacity)
to manage complex economic systems and perfect long-range planning.
However, a comprehensive reconceptualization of the ways in which
technological, organizational, and ideological changes are intertwined
with one another and with the broader socioeconomic context has
remained elusive.

This book attempts such a reconceptualization, deriving a new
theory of work organization and organizational control from the ample
empirical documentation of the past few decades. Organizational control structures include structural characteristics and corresponding ideologies that work together to insure managerial control of the labor

1

process, subordination of the work force, and legitimation of this sub-ordination.[1] In recent years, a new type of organizational control structure has become apparent in production workplaces, corporations, and professional organizations, which I propose to call *technocracy*.[2]

Different forms of organizational control have evolved through a dialectical process of rationalization: precapitalist craft/guild control, simple control, structural forms of control (technical control, bureaucracy, professionalism), and technocratic control. Technocracy is a synthetic type of organizational control, one that integrates certain aspects of the previous forms of structural control: technical control, bureaucracy, and professionalism. Although technocratic control currently coexists with earlier forms of organizational control, it has become sufficiently prevalent to allow comprehensive analysis.

Technocratic organization is currently most apparent in workplaces centered around computerized technology: highly automated production workplaces, "high-tech" research and development corporations, and service corporations that are heavily reliant on computerized systems, such as telecommunications and finance. As certain factories, bureaucracies, and professional organizations are restructured around advanced technology and technical experts, similar changes in structure have become apparent. The central features of technocratic organization include a polarization into expert and nonexpert sectors,[3] a flattening of bureaucratic hierarchies, an erosion of internal job ladders and increased emphasis on credentialing and credential barriers, increased salience of technical expertise as the primary source of legitimate authority, and flexible configurations of centralization/decentralization. (See chapter 6 for a fuller discussion of these features of technocratic organizations.)

PRIOR DEFINITIONS OF TECHNOCRACY

My usage of the term *technocracy* is a departure from previous definitions. The most general definition of technocracy has been "rule by experts," and given the increased salience and centrality of experts within work organizations, this usage is not inconsistent with my own, although technocracies have other important structural characteristics, and the rule of the experts is by no means absolute or unproblematic, as we shall see.

Some have used the term *technocracy* to imply not only the power of experts, but also the desirability and/or inevitability of such rule. Such defenses of expert rule go back at least to Saint-Simon and became the basis for a political movement in the 1920s and 1930s (see chapter 2). The general line of argument here is that, given the increasingly scientific nature of advanced industrial society, only those with technical expertise are qualified to govern and manage, and such rule would enhance meritocracy and industrial productivity.[4] This usage of *technocracy* is diametrically opposed to my own, as I believe both that the evidence challenges the view that technocracy implies meritocracy, and that the polarization into expert and nonexpert sectors within work organizations raises more problems than it solves.

Technocracy has also been used to imply technological determinism, in that advanced industrial technology is viewed as a powerful independent variable, shaping society in such a way that technical experts must be in positions of power.[5] I am not a technological determinist, although I do emphasize technological change as *one* important causal variable.[6] I see technology as embedded in a nexus of other variables: political, economic, and social variables that shape the development of technology and how a given technology is designed and implemented, even as the form of technological systems reciprocally affects these variables. My approach is therefore more dialectical, emphasizing embeddedness and reciprocal causation. However, I do see technological change as one dynamic and important dimension of organizational change, and one with real effects. The design and implementation of technology are shaped by many social variables, but once designed and implemented, the technology becomes an important causal variable, and one with lasting effects; technology is therefore *flexible* but not *neutral*.

The final connotation of technocracy that must be discussed is the idea that technocracy implies that technical considerations have displaced political ones, effectively rendering politics obsolete. This, in fact, is the central thrust of technocratic ideology: the idea that there is "one best way" to accomplish any task, that this way can be found and implemented only by technical experts, and that political concerns are therefore no longer operative. Conversely, I argue that technical considerations interact with political ones in diverse ways, and that politics, although different in technocratic systems, is by no means obsolete.

Given the fact that *technocracy* is such a value-laden term, the question arises: Why use it at all? I use it first of all to suggest that we have experienced a fundamental break with bureaucratic organization, that we need a new theoretical paradigm in order to understand contemporary work organizations. Second, the various connotations and previous definitions, while often ideological in their most extreme versions, are not without a grain of truth: Technocracy does involve a new centrality of experts, the increased importance of technology as a causal variable, and a complex interweaving of technical and political concerns. Only through a more thorough and sustained analysis of contemporary changes in work organizations can we go beyond the partial (and hence ideological) definitions of the past and achieve a more comprehensive understanding.

THE DIALECTICAL RATIONALIZATION OF ORGANIZATIONAL CONTROL

Because technocratic organization and ideology are synthetic forms, adequate understanding of them is dependent on historical analysis of the evolution of organizational control. Organizational control structures have evolved through a dialectical process of rationalization. A rough chronological sequence can be traced, although uneven development and the coexistence of different forms of control is the norm. The different control structures, from precapitalist craft/guild control, to simple control, to structural forms of control (technical control, bureaucracy, professionalism), and finally to technocratic control, have superseded one another as the contradictions within and between control structures have become manifest, leading to crises that were addressed with alternative forms of control.

Organizational ideology plays a central role in this pattern of dialectical rationalization as well. Michael Urban defines organizational ideology in the following way:

> The question of a struggle over power and position within human organizations raises with it the issue of ideology . . . understood here as a more or less coherent set of ideas generated by conditions of conflict or contradiction obtaining in society. Such ideas reflect this conflict, but in a refracted manner, so as to render it resolvable at the level of symbols. Consequently, ideologies simultaneously reveal and conceal something about the conditions in which they are born, and by concealing, they tend to perpetuate these conditions.[7]

Ideologies are inherently related to the types of conflict and contradiction emerging from different types of societal and organizational inequality, and they have changed in accordance with the development of different forms of organizational control. Technocratic control is only the most recent stage of rationalization and continues to embody contradictions that point toward further change. Table 1 provides a summary of the discussion of the evolution of organizational control structures that follows.[8]

Precapitalist Control and Ideology

Precapitalist workplace control centered around family-based production and craft/guild control, both of which were buttressed by ideologies of traditionalism and religion. Theocratic ideology promoted powerful ideologies of the necessity of occupational inequality, submission to "the authorities, particularly the Highest authority,"[9] patriarchal domination, and hard work in one's "calling," however lowly.[10] Traditional ideology held that "the rich should be *in loco parentis* to the poor, guiding and restraining them like children,"[11] that paternalism should include noblesse oblige and responsibility for the poor, that work should be performed in accordance with norms of craftsmanship, and that poverty was useful in that it encouraged industriousness.

The stringency of traditional ideology made possible considerable organizational flexibility within families and guilds: The hierarchical relationship among masters, journeymen, and apprentices, like the division of labor within the family, was relatively collegial and fluid.[12] Skilled craftsmen were exclusionary and elitist toward unskilled workers outside the trade, but the actual craft labor process, although personalized and variable, tended to be collegial. Within the family, production was organized according to a clear division of labor according to sex, but these sex-role boundaries were seen as practical rather than psychological in nature and were routinely crossed when necessary.[13] The ideological framework, however, was quite stringent and specific about one's place in the hierarchy and the moral imperative to work diligently, making role transgressions less threatening to the status quo.

Simple Control and Ideology

With the emergence of capitalism, precapitalist organization of work was perceived as insufficiently stringent in terms of control of both the work process and the product.[14] Transitional forms of work organization served to ease the abrupt shift from home-based production to factory

TABLE 1
Organizational Control Structures

Control structure	Period (approx.)	Characteristics	Contradictions
Craft/guild, family production	Precapitalist (pre-18th C.)	-Apprenticeships -Decentralization -Fluid, collegial -Theocratic ideology of gender, class inequality	-Labor process and product insufficiently controlled
Simple control	18th C. to present	-Direct supervision -Coercive authority -Time discipline	-Transparent coercion -Worker resistance -Impractical in large enterprises
Technical control	19th C. to present	-Control embedded in machine system -Machine sets pace -Worker isolation -Deskilling	-Visible pacing, control -Worker resistance -Inflexibility of production systems
Bureaucratic control	19th C. to present	-Differentiation of structure -Specialized job tasks -Promotion from within based on objective criteria	-Professional/bureaucratic conflicts -Favoritism vs. alleged objectivity -Inefficiency and inflexibility with atypical cases
Professional control	19th C. to present	-Status groups -Self-regulation -Ethical codes -Formalization of training -Esoteric skills	-Professional/bureaucratic conflicts -Vested interests vs. ethics, self-regulation
Technocratic control	1960s to present	-Polarization into expert and nonexpert -Erosion of internal job ladders -Technical expertise as basis of authority -Credential barriers -Team organization, mostly in expert sector -Skill restructuring -Centralization/ decentralization -Ideology of technical imperatives and system maintenance	-Productive potential of advanced technology thwarted -Increased reliance on nonexpert workers -Alleged neutrality vs. race and sex segregation

life: the putting-out system and the system of internal subcontracting.[15] Workers had to shift from working in accordance with natural rhythms, an emphasis on completion of tasks, and alternating periods of industry and idleness to time discipline and regularized work schedules—a major shift in work habits and work culture.[16]

The factory system ultimately relied on simple, direct control and its corresponding ideology of capitalist ownership as conferring control prerogatives. Both entrepreneurial control of entire enterprises and the control exercised by foremen were legitimated by this ideology of property rights:

> Hierarchical control was based on the concept that each boss—whether a foreman, supervisor, or manager—would re-create in his shop the situation of the capitalist under entrepreneurial control. . . . "the foreman's empire." Each boss would have full rights to fire and hire, intervene in production, direct workers as to what to do and what not to do, evaluate and promote or demote, discipline workers, arrange rewards, and so on; in short, each boss would be able to act in the same arbitrary, idiosyncratic, unencumbered way that entrepreneurs had acted.[17]

Simple control would therefore not have been possible without the corresponding legitimating ideology of entrepreneurial prerogative.

Storey discusses this early entrepreneurial ideology as comprised of several aspects.[18] First, ownership rights and responsibility. Second, the belief that "there are persons naturally identifiable as 'leaders,' and others who perform best when led,"[19] an outgrowth of social Darwinist ideas of the survival of the fittest. Given the Darwinian struggle for capital, capitalists were thought to have proven themselves to be superior individuals by accumulating capital. Third, the idea that because of this natural superiority and ability to lead, capitalists can best serve the *general* interest by exercising workplace control and making their enterprises profitable.[20]

As the size of capitalist enterprises grew and the hierarchical distance between owner and supervisor increased, simple control became less effective, because the overt coercion of foremen and direct supervisors was transparent and poorly legitimated by entrepreneurial ideology. By the midnineteenth century, then, industrialization and the consolidation of capital created both structural and ideological changes and a "crisis of control."[21] Three separate types of control emerged as alternatives and adjuncts to simple control: technical control (in blue-collar production workplaces), bureaucratic control (in white-collar, corporate settings), and professional control (in nonroutinized, skilled work settings). All three of these organizational innovations represent more *structural* types of control, as compared with the personalized supervision characteristic of simple (and precapitalist) control: "Rather than being exercised openly by the foreman or supervisor, *power was made invisible in the structure of work.*"[22]

Technical Control and Ideology

Technical control is embedded in the design of machines and mechanical systems, so that they set the pace and form of work. In contrast to the overt coercion characteristic of simple control, workers are constrained by the design of the machine technology to work at a certain pace and in specific ways. Although the technology is designed and implemented by industrial engineers with political as well as technical motives in mind, control motives are less apparent to workers because they are not involved in the design process, and the technology thus tends to appear to them as a fait accompli.

The clearest manifestation of technical control is continuous-flow production, for instance, as used by the early textile industry and (most famously) the automobile assembly line. Early forms of mechanization typically relied more on simple control than on technical control, but as worker challenges to simple control became more frequent and the technology more sophisticated, mechanized assembly lines became increasingly common by the late nineteenth and early twentieth centuries. Driven by the desire to increase efficiency in order to meet consumer demand, Ford created an assembly line that perfected earlier forms and dramatically reduced the need for foremen. In diverse workplaces, workers were deskilled and isolated, as craft skills were rapidly eroded.[23]

Technical control depends on three corresponding legitimating ideologies: technological autonomy/neutrality, technological determinism, and technological progress. Workplace technology, like technology in general, is presented as driven by its own dynamic rather than by particular interests. As Alvin Gouldner put it: "What 'technology' does is to present itself as a universal, all-purpose praxis, as a practice fit for the pursuit of any and all goals and as available to all and every group."[24] The particular form that technology takes is therefore assumed to be a function of the general state of technological progress; to challenge technology is assumed to be irrational, as contrary to both technological and societal progress. Technological advances are assumed to be inherently progressive due to a "cult of productivity and expertise."[25] Technological choices have been presented as the inevitable manifestations of science/engineering and as inherently humane and liberating. Typically, technological innovation *does* involve some improvements in efficiency and performance, a fact that has contributed to the obscuring of technical control as well as the decreased likelihood of conceptualizing alternative forms of technological design or implementation.

Technical control was not without its contradictions, however. Although control was less personalized than simple control, the workplace technology was nonetheless concrete and visible, and certain aspects of technical control were quite transparent (e.g., the variable speed of the assembly line). In some workplaces, the contradiction between the expense of, and care given to, the technology, and the low wages of, and disregard for, workers, was pronounced. Moreover, because the technology was both expensive and deskilling, the ongoing reliance on deskilled and often disgruntled workers became problematic for some owners.[26] Technical control eroded craft traditions and created a more homogeneous group of semiskilled workers, which promoted solidarity among workers but reduced worker motivation from the capitalist point of view. Unionization and other forms of resistance among workers were common.[27] In response to these limitations of technical control, owners turned to auxiliary, more bureaucratic, forms of control such as the development of internal job ladders (despite the homogenization of skill) and piece rate systems.[28]

Bureaucratic Control and Ideology

In contrast to technical control, bureaucracy rests on an expansion of formal/legal rationality within work organizations: the reliance on specified formal rules and their structural manifestations. Bureaucracy structures white-collar, administrative work in ways analogous to the ways in which mechanization structures material production. James Beniger argues that industrialization initiated control crises in other domains of society, particularly administration, distribution, and communication, and that bureaucracy emerged as a "critical new machinery . . . for control of the societal forces unleashed by the Industrial Revolution."[29]

Unlike the personalized work culture of both craft work and simple control, bureaucratic rules purport to constrain workers at all levels of the organization, even as they protect workers from arbitrary exercises of power.[30] Moreover, bureaucratic procedures streamline information processing by reducing the amount of relevant information to delimited "cases" and by moving "from the government of men to the administration of things." Such rationalization makes possible the emergence of larger and more complex social systems.[31]

Bureaucratic organization involves task specification and differentiation, a hierarchy of offices separate from their incumbents, promotion upward through the ranks, centralized and specified authority channels, and objective criteria for evaluating promotion and remuneration.[32] In a bureaucracy, the division of labor is expanded to include not only production tasks but also supervisory/administrative tasks; clearly delimited spheres of responsibility and distinct supervisory levels are part of the effort to rationalize administration. The pyramidal structuring of these layers and a body of rules governing workplace functioning became characteristic of burgeoning corporations during the late nineteenth century.

Bureaucratic ideologies are several. As with technical control, bureaucracy is legitimated on the grounds of impartiality and neutrality: clearly defined rules that apply to everyone, allowing all to compete equally and fairly. Moreover, as Reinhard Bendix points out, the promotion from within upward through the ranks is related to the ideology of "rags to riches," a central ideology of capitalist culture.[33] Seniority and competence are assumed to lead to elevated rank authority, so that the bureaucratic organization will be meritocratic. Like technical control, bureaucracies are also presumed to increase efficiency and productivity.

Managerial ideology is another way in which bureaucratic power is legitimated. Managerialism emphasizes the managerial function as separate from and more important than any other job in the bureaucracy, as a difficult art that can be performed only by certain types of individuals who have "rare qualities . . . and unique competence."[34] In addition to the formal rationality of bureaucracy, certain charismatic qualities of leadership are emphasized: "Rather than only specifying rules and regulations to govern various work situations, managerial ideologies function to promote an atmosphere or attitude of loyalty."[35] Moreover, managers are assumed to be altruistic, working for the good of the organization rather than for self-interest.

As with earlier forms of capitalist control, dysfunctions and contradictions of bureaucracy have emerged. The relative inflexibility of bureaucratic structure has implied inefficiency with atypical cases. As size and complexity have led to more layers of bureaucratic organizations, communication between the top and the bottom has become problematic.[36] Despite the ideology of objectivity and neutrality, favoritism and personalized identification with positions are common. The personalized nature of managerial ideology implies a corresponding vulnerability to perceptions of personal failure. Due to the emphasis on seniority, promotion from within, and centralized rank authority, technical experts and professionals have resided uneasily within bureaucratic organizations, leading to "professional/bureaucratic conflicts" and a less innovative work organization.[37] Both worker and client resistance to these limitations of bureaucracy have been common.

Professional Control and Ideology

Professionalism arose simultaneously with technical and bureaucratic control during the latter half of the nineteenth century, but largely in reaction to these alternative forms of structural control. Professional control differs from technical and bureaucratic control in that it allows for more worker discretion in dealing with clients and more autonomous and collegial forms of work organization. Professionals are alleged to possess esoteric skill and knowledge, which necessitates self-regulation and collegial control rather than the external control characteristic of bureaucracy or technical control.[38] In direct contrast to the bureaucratic processing of delimited cases, professionals are expected to deal with clients as unique individuals, utilizing formalized knowledge and skill to formulate professional judgments about personalistic situations. Indeed, professional work is often viewed as the antithesis of bureaucracy, as creating "professional/bureaucratic conflict" unless alternative forms of work organization are found (see below).

Professional control has two main dimensions: collegial self-regulation and professional control of client relations.[39] Collegial self-regulation centers around the formalization of professional training (including entrance requirements, curriculum, exit requirements, and credentialing) and the monitoring of professional conduct through collegial organization and professional associations.[40] Another aspect of self-regulation concerns norms of service and professional conduct: objective, impersonal, nondiscriminatory, and quality work, guided by the service ideal. Such norms, although usually embodied in codes of ethics, are largely enforced informally.[41] These norms and codes of ethics also influence professional/client relations by creating social distance and respect. Professional/client relations are also influenced by the broad-based nature of professional competence—esoteric and yet practical, involving both formalized and tacit knowledge—and the typical context of crisis or client need—both of which imply status inequality and professional control of the situation.[42]

Professionalism has certain affinities with craft work: conceptions of work as intrinsically interesting and valuable, collegial working conditions, antimarket ethics of "good work" and a general service orientation, and noblesse oblige, or the sense that professional privilege implies duties and norms of behavior.[43] Unlike traditional craft work, however, professional training has become more and more formalized and institutionalized, and professional knowledge has become commodified into a "special kind of property."[44] Legitimation of the professional credentialing process (and of professional competence in general) is more and more dependent upon the operation of the educational system and, in particular, the degree of perceived equality of educational opportunity. The structural and hidden aspect of professional control concerns the process of professional training and certification; the degree to which professionals are accorded legitimacy rests to a large extent on the perceived legitimacy of this system of professional training and credentialing: "The university is the center from which ideological legitimation radiates."[45]

Professional ideology, then, rests on client perceptions of the legitimacy of the selection and training process, as well as generalized beliefs concerning the degree of efficacy of professional codes of ethics, particularly norms of service and absence of self-interest. Skepticism regarding professional ideology has been commonplace, however: the sense that self-interest and mercenary motives are not incompatible with professional codes of ethics, that professional review boards are often insufficient to provide adequate regulation, and that professional privilege is derived not only from the nature of professional work but also from professional interest-group politics.[46]

The Contradictions of Control

As we have seen, all three forms of structural control were from the beginning plagued by contradictions of control and by worker/client resistance. Because two or more control structures typically coexist, there are also contradictions *between* these various control structures.

The persistence of internal subcontracting within factories in some industries (e.g., steel, coal mining, machining) implied an uneasy coexistence of craft control and simple control. The polarization into workers versus management, which is pronounced in workplaces under technical control, implies reduced or nonexistent mobility prospects for workers, which contradicts the ethos of individual opportunity and upward mobility, characteristic of both professional and bureaucratic models and of American society in general. In enterprises that combine a technical production sector with a more bureaucratic administrative sector, the different mobility prospects of each sector may be highlighted.

The most fully analyzed set of inter-control-structure contradictions has been "professional/bureaucratic conflicts."[47] Particularly in organizational contexts where professionals are working within a bureaucracy, a situation that has become more prevalent during the last century, the autonomy and self-regulation characteristic of professional control can come into contradiction with bureaucratic rules and authority structures. The central contradiction here is between authority based on knowledge and authority based on seniority and rank position, which sets the stage for conflict over the locus of authority. A corollary is that professionals working within bureaucracies may find that their allegiance to their profession can conflict with their loyalty to the organization. Moreover, the bureaucratic ethos of efficiency and corporate profit maximization can contradict the professional ethos of competence, status, and client orientation.

One way in which these intra- and interorganizational contradictions were initially addressed was through the attempt to professionalize management and administration so as to rationalize factories, bureaucracies, and professional organizations. Taylorism and its search for the scientific basis of management was an early attempt to increase managerial control and organizational efficiency by expanding management's technical understanding of the labor process.[48] Professional managers were assumed to occupy a stronger position vis-à-vis both workers (whose craft knowledge had been appropriated) and professionals (who were presumed to accord more respect to fellow professionals). Although Taylorism relied on only the most rudimentary science and technical understanding (time and motion studies, combined with cost accounting), the attempt to rationalize and legitimate bureaucratic administration by giving it a surer basis in scientific and technical expertise was an important precursor of technocratic control.

Taylor emphasized the capability of engineering science to discover the "one best way" to solve administrative problems and make production decisions. Diverse workplaces could be rationalized by turning them over to professional engineers, who would insure progress in the form of material development, greater efficiency, and technological advance. However, although Taylorists brought religious fervor, as well as certain technical innovations,[49] to their reform efforts, the scientific basis of scientific management was too weak to provide sufficient rationalization or legitimation. Time and motion studies were conducted publicly, and assembly lines were overtly speeded up—both of which were transparent control strategies. Worker resistance, particularly from the craft unions and professionals, was considerable.[50]

One way in which the contradictions of Taylorism were initially dealt with was through the human relations approach. As Frank Fischer put it:

> From management's point of view, Taylorites had properly identified the issue of workplace authority, but as Richard Edwards put it, they "had not found quite the right mechanism." The human relations movement can be understood as the culmination of a series of interrelated attempts to find the "right mechanism." Early interest in human relations by industrialists can, in fact, be interpreted as a response to the upsurge of organized labor, significantly facilitated by hostilities toward Taylorism itself.[51]

Human relations approaches therefore sought to supplement Taylorist policies with psychological techniques that acknowledged the importance of primary work groups and workers' feelings and motivation. As an ideology, human relations promoted a more insidious form of managerial control, because it obscured this control: Management is presented in paternalistic terms, and the work organization as "one big family."[52]

The search for the one best way to organize production so as to minimize worker/management conflict and increase productivity has grown more intense in recent years, as the scientific and technological basis of workplace rationalization has expanded. In recent years we have seen a "transition from scientific management to the scientific-technical revolution."[53] As David Stark points out, the long-term significance of Taylorism was to begin to legitimate and institutionalize a new class division between manual and mental workers.

> The period of the transformation of the labor process during the early decades of this century was also an important period of class formation

with significant consequences for the contemporary constellation of class relations. . . .What the reorganization of work did accomplish was to provide the basic conditions for the ideologically sharp division between "mental" and "manual" labor. . . . The creed of specialized knowledge and expertise became the formative basis of a new and more complex ideology around which a class could cohere.[54]

Technocratic control is, in effect, a more sophisticated contemporary alternative to Taylorism. During the post-World War II period, and particularly since the 1960s, a set of interrelated socioeconomic changes has fueled the quest for more extensive workplace rationalization. Rising educational levels and more intense competition for commensurate jobs, equal employment opportunity pressures, the development of computerized technology and advanced information systems, the increasing size of the state sector, the emergence of multinational corporations and worldwide competition among them, and the increased need to manage the economic system and perfect long-range planning—these developments necessitated and facilitated subsequent rationalization. Certain aspects of professionalism, bureaucracy, and technical control have become integrated into a more complex and heavily legitimated form of technocratic control.

Contemporary Conceptualizations of Changing Work Organizations

Many have attempted to understand and conceptualize the dramatic changes that have occurred in the occupational sector. Just as the Industrial Revolution spawned diverse theories, so the "second industrial divide" has generated varied theories of contemporary socioeconomic change. Both in scholarly publications and in the popular press, new theories of work organizations have been abundant. Although I believe that none of these theories is adequate, many of them are instructive and worth examining.

One salient feature of many recent analyses of changes in work organizations is their optimism, particularly with regard to the impact and potential impact of computerization on the workplace. In the literature on contemporary organizations, for instance, one finds such conceptualizations as "post-bureaucratic organizations," "post-industrial organizations," "postmodern organizations," "adhocracies," etc.[55] While some of these analyses are more cognizant of the complexity of recent changes than others, the general thrust of all of them is toward opti-mism concerning the impact of computerization and related changes in work organizations. In a useful review of this literature, for instance, Wolf Heydebrand concludes that postbureaucratic organizations tend to open up new democratic possibilities by undermining both bureau-cracy and professional dominance, and that these new organizations are characterized by informalism, universalism rather than special inter-ests, loosely coupled subunits, extraorganizational networking, and enhanced corporate culture.[56] In a similar vein, the literature on post-modernist organizations has affirmed "de-differentiation," difference, challenges to binary oppositions and grand narratives, and an emphasis on "pluralism of cultures, communal traditions, ideologies, forms of life or language games."[57]

The sociology-of-work literature in this field often exhibits a sim-ilarly optimistic tone. Larry Hirschhorn, for instance, concludes from his case studies of advanced technological settings that successful comput-erization depends upon expanding and enhancing workers' skill and knowledge: that workers need a more comprehensive understanding of workplace operations in order to be able to effectively monitor com-puterized operations.[58] In Hirschhorn's view, this necessary expansion of knowledge leads to very beneficial effects:

> As knowledge is incorporated into machines, workers can reinvolve themselves at a wider and more comprehensive level of production. Through a developmental process, machines and workers together increase the store of practical and theoretical knowledge. . . . As we move from preindustrial to postindustrial conditions, the relationship between work and consciousness is dramatically transformed . . . the worker becomes more aware of his work environment, but he also begins to reflect self-consciously on his own actions and becomes aware of how he learns and develops.[59]

For Hirschhorn, this expansion of worker knowledge and conscious-ness implies a corresponding expansion of worker power and control in contemporary workplaces. Fred Block has built on this postindustrial optimism concerning the impact of computerized technology, show-ing how technological change has created the preconditions for gen-uine workplace democracy.[60]

Shoshana Zuboff reaches similarly optimistic conclusions about the *potential* impact of "the smart machine."[61] For Zuboff, computeri-zation points toward a duality: The technology can either automate or informate. By informate she means that the technology can be used to generate and disseminate information about underlying administrative and production processes. When the informating capacity is realized, workers gain more abstract and comprehensive knowledge of the workplace. However, according to Zuboff, whether the informating potential is realized depends upon managerial choices: "Managers can choose to exploit the emergent informating capacity and explore the organizational innovations required to sustain and develop it. Alternatively, they can choose to ignore or suppress the informating process."[62] For Zuboff, however, unless informating is encouraged, the potential benefits of computerized technology cannot be realized.

Rosabeth Kanter's analysis of the organizational prerequisites of innovation expresses a similar duality between organizations that are progressive, both technically and socially, and organizations that are more traditional: *integrative* versus *segmentalist* organizations.[63] The integrative organizations, which tended to be high-tech workplaces, were characterized by a more holistic approach to problem solving, utilization of a team structure, a more cooperative style, less specialization, a matrix style of organization, and a general flexibility that encourages innovation. In a more recent analysis, she discusses "post-bureaucratic" organizations as being more centered on the individual worker and his or her expertise, as "results oriented" rather than rule oriented, as focused on creativity and innovation, and as emphasizing fluid groupings and turnover as positive renewal of expertise and synergistic creativity.[64]

In contrast to these generally optimistic accounts of the effects of computerization and related socioeconomic changes, the neo-Marxist work on computerization has focused on the negative effects of technological change.[65] The general thrust of neo-Marxist work on computerization has been to emphasize its capacity to enhance control of the labor of the nonexpert sector, diminishing autonomy, deskilling the actual labor process, and creating an "electronic sweatshop."[66] Professional work is also seen as being adversely affected by computerization: as "proletarianized."[67] We will examine some of the evidence for these trends in chapters 3 and 4.

Some social scientists have attempted to conceptualize more macrosocial and global political and economic changes. In an influential work, Michael Piore and Charles Sabel have analyzed the "second industrial divide" that computerization represents, and the possibilities for socioeconomic change that it affords.[68] Essentially they see the

options as either a continuation of Fordism (mass production techniques) and international Keynesianism, (techniques more appropriate
to the technology of the industrial revolution) or a return to "flexible
specialization" and craft methods of production, methods that Piore
and Sabel see not only as possibilities, but as "essential to prosperity."[69]
Computerization is seen as promoting flexible specialization:

> The connection . . . between flexibility and computers is supported by
> ethnographic studies of computer users—ranging from schoolchildren
> to sophisticated programmers. . . . Whereas most machines have an inde-
> pendent structure to which the user must conform, the fascination of the
> computer . . . is that the user can adapt it to his or her own purposes and
> habits of thought. The computer is thus a machine that meets Marx's
> own definition of an artisan's tool: it is an instrument that responds to and
> extends the productive capacities of the user . . . technology has ended the
> dominion of specialized machines over un- and semi-skilled workers,
> and redirected progress down the path of craft production. The advent of
> the computer restores human control over the production process;
> machinery again is subordinated to the operator.[70]

Piore and Sabel argue that within the international division of labor
"the old mass-production techniques might migrate to the underdeveloped world, leaving behind in the industrialized world the high-tech
industries . . . all revitalized through the fusion of traditional skills and
high technologies."[71]

Another recent analysis of global economic trends that has been
widely influential is Robert Reich's *The Work of Nations*.[72] For Reich,
global power has become diffused, and once-powerful multinational
corporations have evolved into a web of decentralized international
groups. Given the complexities of the international division of labor
and foreign ownership, it is no longer meaningful to speak in terms of
national corporations; "American" corporations are no longer American
in any meaningful sense of the word. Given this fact, it is highly unre-
alistic to expect that corporate profits will "trickle down" to everyone in
a given nation; Reich argues that the actual trend has been for economic
assets to "trickle out" to whichever global investment seemed most
profitable.[73] Capitalism has become highly "disorganized,"[74] if indeed
we can speak of capitalism at all.

Reich goes on to discuss how the militaristic bureaucracies and
Fordist production techniques of the 1950s have been superseded by
"high-value" firms that emphasize specialized knowledge and innova-

tive ideas. High-value firms cannot be organized bureaucratically, but rather utilize creative teams (including international coalitions), horizontal integration, and high rewards for those who contribute innovative ideas to the global competition. Reich sees three main categories of workers: routine production workers, in-person service workers, and symbolic analysts.[75] Production workers are diminishing; in-person service workers are increasingly in demand but are poorly paid; and symbolic analysts are in an increasingly powerful position, to the extent that their material and nonmaterial rewards are leading to their veritable *secession* from their countries of origin.

What are we to make of these varied conceptualizations? One striking fact is that many of them, particularly those in the postindustrialist tradition, are highly optimistic scenarios of the types of organizational and socioeconomic changes that computerization is facilitating. This optimism is even more apparent in the journalistic media, where a type of technological hubris implies that technological innovation can solve virtually any social problem. In these accounts, technological determinism becomes liberation through technology.

At the opposite extreme, we find those accounts, many inspired by neo-Marxism, that stress the dangers of technological change: enhanced managerial control, deskilling of the labor process, new types of stress and occupational health problems, technological unemployment. The empirical evidence underlying these analyses indicates that working with computers is more oppressive than liberating, although many of these writers see the potential for technology to be implemented in a more humane manner.

Some of the more nuanced analyses have avoided the pitfalls of technological determinism and have sought to clarify the choices that have accompanied computerization: to automate or informate the labor process, Fordism or flexible specialization, integrative or segmentalist organizations. What has generally been lacking in these analyses, however, has been an awareness of the fact that these polar extremes currently coexist within work organizations. Most of these writers have also assumed that the technological imperative is on the side of the more progressive option: that in order for the benefits of computerization to be realized, positive organizational innovation must occur.

These, then, are the perspectives on contemporary socioeconomic reality from which I have learned the most. To go beyond them is an ambitious task. However, I believe that we must transcend the Manichaean dichotomizing of technology and its effects: Computerization is neither good nor bad, but a complex tool that has the potential to be implemented in diverse ways. Moreover, technological change is embedded in a nexus of social and political variables and must be analyzed in conjunction with them. Only by analyzing, in some detail, the actual patterns of organizational restructuring in various workplaces and their effects on workers and workplaces can we achieve the type of comprehensive reconceptualization that will enable us to make liberating technological and organizational choices for the twenty-first century.

SYNOPSIS OF CHAPTERS

Chapter 2 explores the intellectual history of the concept of technocracy, including its source in Enlightenment theory, the technocratic movement of the 1930s, postindustrialism, new class theories, and postmodernism. Given the fact that technocracy is a synthetic organizational form, derived from a combination of certain aspects of technical, bureaucratic, and professional control, chapters 3 through 5 explore the dynamic changes in blue-collar, white-collar and professional work so as to understand the basis of technocratic control. Chapter 3 analyzes the changing nature of technical control by reviewing recent empirical studies of blue-collar production workplaces and how they have changed as a result of computerization and related socioeconomic developments. Chapter 4 examines the recent literature on bureaucracies, and chapter 5 looks at recent studies of various professions. Chapter 6 extrapolates from chapters 3 through 5 to set forth a theory of emergent technocratic organization and how it differs from previous forms of control. Chapter 7 concludes by looking at the political significance of technocratic organization and control and prospects for further changes in work organizations.

Chapter 2

THE IDEOLOGICAL ROOTS OF TECHNOCRACY

> *Most of us are conditioned for many years to have a political viewpoint, Republican or Democratic—liberal, conservative, moderate. The fact of the matter is that most of the problems . . . that we now face are technical problems, administrative problems. They are very sophisticated judgments which do not lend themselves to the great sort of passionate movements which stirred this country so often in the past. Now they deal with questions which are beyond the comprehension of most men, most governmental administrators, over which experts may differ.*
> —John F. Kennedy, 1962 (quoted in Frank Fischer, Technocracy and the Politics of Expertise)

Technocracy is a concept with a long and rich ideological history. Even though we have defined technocratic organization rather specifically and narrowly, a more general consideration of the evolution of the concept of technocracy is in order. Theorists from diverse orientations have addressed issues concerning experts, expert knowledge, and related social and political implications. Their differing definitions and interpretations constitute an intellectual history that is of continuing relevance to contemporary technocracy, in particular because prior conceptualizations of experts and expertise continue to influence societal perceptions of how work should be organized.

FROM THE ENLIGHTENMENT TO TECHNOCRACY, INC.

The roots of technocracy go back to the Enlightenment, with its emphasis on reason, science, technical rationality, and technology. A central theorist of the new Enlightenment ethos was Francis Bacon. Frank Fischer describes the shifting worldview that Bacon promoted in the following way:

> For Bacon, the defining feature of history was rapidly becoming the rise and growth of science and technology. Where Plato had envisioned a society governed by "Philosopher kings," men who could perceive the "forms" of social justice, Bacon sought a technical elite who would rule in the name of efficiency and technical order. Indeed, Bacon's purpose in *The New Atlantis* (1622) was an explicit attempt to replace the philosopher with the research scientist as the ruler of the utopian future. New Atlantis was a pure technocratic society.[1]

Bacon's vision of what a technocratic society could be emphasized hierarchy and patriarchy; in fact, the ruling scientists were called the "Fathers," and women not only were excluded from positions of power but were "reduced to near invisibility."[2] The ruling scientific elite was seen as enlightened and benevolent, ruling in the general interest:

> In *The New Atlantis*, politics was replaced by scientific administration. No real political process existed in Bensalem. Decisions were made for the good of the whole by the scientists, whose judgment was to be trusted implicitly, for they alone possessed the secrets of nature. . . .
>
> The scientist father was portrayed much like the high priest of the occult arts, the Neoplatonic magus, whose interest in control and power over nature had strongly influenced Bacon. He was clothed in all the majesty of a priest, complete with a "robe of fine black cloth with wide sleeves and a cape. . . ." His gloves were set with stone, his shoes were of peach-colored velvet, and he wore a Spanish helmet. . . .
>
> Bacon's scientist not only looked but behaved like a priest who had the power of absolving all human misery through science. He . . . "held up his bare hand as he went, as blessing the people, but in silence." The street was lined with people who, it would seem, were happy, orderly, and completely passive.[3]

Saint-Simon continued and elaborated further on Bacon's vision of a new type of society, one organized around scientific rationality. Writing over a century later, but still very much in the context of the Enlightenment enthusiasm for scientific method as expressed by Bacon, Descartes, and Newton, Saint-Simon too saw science as an appropriate modern replacement for religion, at least among the educated. Central to Saint-Simon's political agenda was his vision of a ruling elite composed of scientists, artists, and industrialists; he wrote in 1803:

> I believe that all classes of society would benefit from an organization on these lines: the spiritual power in the hands of the scientists, the temporal power in the hands of the property-owners; the power to nominate those who should perform the functions of the leaders of humanity, in the hands of all; the reward of the rulers, esteem.[4]

For Saint-Simon, the central reason why such an elite should have power in society was a functional one: Scientists, artists, and artisans were the productive and capable members of society. In some of his essays, property owners and industrialists are also included as members of the "enlightened elite." In 1819, he posed the question of what would happen to France if her best scientists, artists, physicians, bankers, and artisans were to be "lost"; his answer was that France would become a "lifeless corpse"; conversely, if France were to lose key members of its aristocracy, "it would be very easy to fill the vacancies which would be made available."[5] For Saint-Simon, the unfortunate reality was that the educated elite were generally subordinate to those of rank authority.

> The scientists, artists, and artisans, the only men whose work is of positive utility to society, and cost it practically nothing, are kept down by the princes and other rulers who are simply more or less incapable bureaucrats. Those who control honours and other national awards owe, in general, the supremacy they enjoy, to the accident of birth, to flattery, intrigue and other dubious methods. . . .
> Society is a world which is upside down . . . in every sphere men of greater ability are subject to the control of men who are incapable."[6]

For Saint-Simon, the solution was to rebuild society along more rational and scientific lines, with the most educated and functional members of society in positions of power and the idle and the uneducated relegated to subordinate positions. The general rule of this social restructuring should be: "In the general interest, domination should be proportionate to enlightenment."[7] That such an educated and enlightened elite would rule benevolently he never doubted. Indeed, his faith in science and "positive" knowledge included the idea that it would effectively transcend politics: "The philosophy of the eighteenth century has been critical and revolutionary: that of the nineteenth century will be inventive and constructive."[8] For Saint-Simon, and later for Comte, the rationality of science precluded despotism; as Comte put it: "The fear of a despotism founded on science is a ridiculous fantasy, because the allegiance of the people to their new scientific leaders would be of quite a different character from the unreasoning obedience to priests in the theological phase."[9] Both Saint-Simon and Comte saw the emerging scientific ethos as a unifying ideology; class struggle and politics would be supplanted by technical decision making.[10]

Frederick W. Taylor assumed the role of conveying technocratic ideas to the United States and giving practical and concrete expression to the concern with societal restructuring along more scientific lines. For Taylor it was the labor process itself that led to the greatest inefficiency and waste, due to the "systematic soldiering" of workers. Taylor advocated scientific management of the work process so as to find the "one best way" to solve any industrial problem:

> Now, among the various methods and implements used in each element of each trade there is always one method and one implement which is quicker and better than any of the rest. And this one best method and best implement can only be discovered or developed through a scientific study and analysis of all of the methods and implements in use, together with accurate, minute, motion and time study. This involves the gradual substitution of science for rule of thumb throughout the mechanic arts.[11]

For Taylor, the one best scientific method is too complex for ordinary workers to understand; therefore, expert management based on scientific principles is necessary. The rather rudimentary scientific principles he advocated included time study, standardization of tools and movements, establishment of a separate planning room, use of time-saving implements (such as slide rules), and use of differential piece rates.[12]

According to Taylor's vision, the scientific management of industry (and ultimately of society as a whole) would reduce production costs and increase efficiency and productivity; it would also lower prices and increase wages.[13] Obviously, in such a scenario, businessmen, engineers, workers, and consumers all stood to benefit. For Taylor and his followers, scientific management became a moral mission, one that was guaranteed to reward the hardworking, punish the lazy, eliminate politics, and restore social harmony. In the private sector, there was considerable interest in Taylorism, coupled with resistance from both managers and workers, with both groups fearing inroads on their autonomy and control.[14] In the public sector, Taylorism was embraced by the Progressive reform movement, which sought to "engineer the transition to a new and 'more rational' form of governance . . . to replace . . . political irrationality with scientifically designed decision processes."[15] A new class of scientific experts was emerging, legitimated on the grounds that they would be above politics and class conflict.

Thorstein Veblen, in *Engineers and the Price System* (1921), was also concerned with issues of social productivity and efficiency.[16] For Veblen, however, it was capitalists and industrialists who were guilty of soldiering. He saw the "captains of industry" as another "leisure class," one that had become as antithetical to social progress as the aristocracy had been for Saint-Simon a century earlier.

For Veblen, industrial society in the 1920s was of a modern and scientific nature:

> In more than one respect the industrial system of today is notably different from anything that has gone before. It is eminently a system, self-balanced and comprehensive; and it is a system of interlocking mechanical processes, rather than of skillful manipulation. It is mechanical rather than manual. It is an organization of mechanical powers and material resources, rather than of skilled craftsmen and tools; although the skilled workmen and tools are also an indispensable part of its comprehensive mechanism. It is of an impersonal nature, after the fashion of the material sciences, on which it constantly draws. It runs to "quantity production" of specialized and standardized goods and services. For all these reasons it lends itself to systematic control under the direction of industrial experts, skilled technologists, who may be called "production engineers," for want of a better term."[17]

The problem for Veblen, as for Saint-Simon, was that the capable, educated, and productive members of society, the production engineers, were subordinated to an obsolete and inappropriately educated group: for Veblen, the capitalist captains of industry. According to Veblen, businessmen and financiers tended to be proficient in the economics of making capitalist enterprises profitable, but unable to maximize industrial productivity:

> Business men are increasingly out of touch with that manner of thinking and those elements of knowledge that go to make up the logic and the relevant facts of the mechanical technology. Addiction to a strict and unremitting valuation of all things in terms of price and profit leaves them, by settled habit, unfit to appreciate those technological facts and

values that can be formulated only in terms of tangible mechanical performance. . . .

They are experts in prices and profits and financial maneuvers; and yet the final discretion in all questions of industrial policy continues to rest in their hands. They are by training and interest captains of finance; and yet, with no competent grasp of the industrial arts, they continue to exercise a plenary discretion as captains of industry. They are unremittingly engaged in a routine of acquisition, in which they habitually reach their ends by a shrewd restriction of output; and yet they continue to be entrusted with the community's industrial welfare, which calls for maximum production."[18]

Indeed, for Veblen, to entrust the technologically advanced industrial system to capitalist businessmen was inevitably counterproductive, as the logic of capitalism required what he termed "sabotage" in order to restrict production and avoid periodic crises of overproduction. Thus, Veblen agreed with Taylor that productivity was being restricted, but disagreed as to the source of the restriction.

Veblen prophetically wrote in 1921 that "the date may not be far distant" when such capitalist "sabotage" would result in a "fatal collapse" of the industrial system.[19] In order to avoid this eventuality, Veblen argued that engineers and technical specialists must replace capitalists as the ruling class of the increasingly complex technological system:

For the due working of this [industrial system] it is essential that that the corps of technological specialists who by training, insight, and interest make up the general staff of industry must have a free hand in the disposal of its available resources, in materials, equipment, and man power, regardless of any national pretentions or any vested interests. Any degree of obstruction, diversion, or withholding of any of the available industrial forces, with a view to the special gain of any nation or any investor, unavoidably brings on a dislocation of the system; which involves a disproportionate lowering of its working efficiency and therefore a disproportionate loss to the whole, and therefore a net loss to all its parts."[20]

For Veblen, to give such power to engineers and technical experts was not only rational, but also democratic. He argued that the technological system was "a joint stock of knowledge and experience held in common by the civilized peoples."[21] The technical experts, upon whom the system's successful operation depends, have been trained at the cost of the community, and the general population's welfare depends upon their being given discretionary power. Veblen argued that the interests of the technicians are fundamentally universal, as opposed to the vested interests of the captains of industry, which are centered around personal gain.

Engineers, then, were the appropriate revolutionary class, well suited to challenge capitalist control. Unlike the traditional working class, engineers and technicians were capable of effectively administering the complex industrial system. In fact, Veblen considered working-class organizations such as trade unions to be yet another vested interest group, "as ready as any other to do battle for its own margin of privilege and profit."[22] Furthermore, because of the adversarial nature of trade unions, they were also implicated in the sabotage of the potential productivity of the industrial system.

Veblen concludes *The Engineers and the Price System* by calling for a "soviet of technicians" to seize power from the capitalist class and direct the industrial system in a more rational and efficient manner. He argues that the engineers and technicians are well placed to seize power, in that they already have significant authority within the system; the problem is that "the businesslike deputies of the absentee owners sagaciously exercise a running veto power over the technicians and their productive industry."[23] However, as a result of their technical superiority, it would be relatively simple for engineers to seize power by means of a general strike; because of their functional importance to the system, they could easily incapacitate it, given sufficient solidarity and will to do so.

Although, as Veblen realized, engineers were not radical enough in their political orientation to make this scenario realistic, his analysis did crystallize and express an ongoing tension between engineers and big business. In the immediate post-World War I period, some engineers were calling for a system of national planning to be coordinated by engineers.[24] Herbert Hoover, an engineer himself, addressed this need in 1920:

> The time has arrived in our national development when we must have a definite national program in the development of our great engineering problems. Our rail and water transport, our water supplies for irrigation, our reclamation, the provision of future fuel resources, all cry out for some broadvisioned national guidance.[25]

In 1919, one engineer, William H. Smyth, coined the term *technocracy* to express his vision of scientific reorganization of society combined with democratic control.[26] Like Veblen, Smyth had a vision of a "National Council of Scientists" that would control "all national institutions."[27] How democratic control was to be assured was not clear; it appears that a group of scientists was considered to be inherently democratic and antithetical to autocracy. Both Smyth and Veblen appear to have merely taken democracy for granted.[28]

By the mid-1920s, Veblen's vision had also been termed "technocracy," by the economist John Clark,[29] and became something of a manifesto for the nascent technocracy movement. Between 1918 and 1921, Veblen had interacted and collaborated with Howard Scott, a radical engineer whose ideas about social engineering closely paralleled his own. As a first step toward social restructuring, they discussed the possibility of a nationwide survey of industry, aimed at documenting the wastefulness of the system and providing the information needed to conceptualize alternatives.

In 1919, Scott formed the Technical Alliance, a group of engineers, scientists, and technicians based in New York. The research project designed by Veblen and Scott was the group's first priority:

> The projected survey would examine 3,000 different industries over the past century and chart their changes in employment, productivity per employee, horsepower capacity, and horsepower used in production. The uniqueness of the project, aside from its imposing scope, was its attempt to measure production and waste in terms of horsepower rather than in the standard categories of labor expended or monetary costs, a clear indication that Scott was working toward an energy theory of value."[30]

During the same period, Scott was working as the research director of the Industrial Workers of the World, where he conducted a series of studies of waste. However, although the aim of these research efforts was to expose the inefficiency of the capitalist system, thereby promoting a more rational and productive type of industrial organization headed by technical experts, the studies failed to achieve this political effect. Although some waste was documented, the prosperity and conservatism of the 1920s undermined the political orientation of the Technical Alliance, and it disbanded by the mid-1920s.

The coming of the depression created more fertile soil for Scott's radical ideas of social engineering. Together with Walter Rautenstrauch, chairman of Columbia University's Department of Industrial Engineering, he formed the Committee on Technocracy in 1932. Rautenstrauch combined progressive politics with a strong belief in the power of engineers. During World War I, as secretary of the American Society of Mechanical Engineers, he had attempted to pressure Woodrow Wilson into adopting scientific management of industry:

> Rautenstrauch tried to impress on Wilson the need "to take control of the huge and delicate apparatus of industry out of the hands of idlers and wastrels and deliver it over to those who understood its operation." This was necessary to achieve the harmony that efficiency and expanded production would create. . . . In the letter to Wilson, Rautenstrauch reiterated Gantt's claim that rational managerialism constituted "true democracy." Whereas parliamentary "democracy accomplishes nothing and leads nowhere," real democracy—which he defined as "the organization of human affairs in harmony with natural laws"—allowed "each individual . . . an equal opportunity to function to his highest capacity."[31]

Rautenstrauch's ideas were broadly consistent with Scott's, and the Committee on Technocracy aimed to begin implementing their program, initially by completing the research project of industry.

By 1933 the Committee on Technocracy was beginning to publish conclusions, even though the research was not completed. These were the ideas that the general public came to associate with technocracy:

> The central thrust of their analysis was the assertion that this new technology, the product of science and engineering, possessed the potential for creating either abundance or a cataclysmic crisis. The inability of society to adjust to the new technology had brought the country to the depression and the brink of disaster. Instead of harmony and abundance, the introduction of high-energy technology had brought waste, inefficiency, pyramiding debt, and most importantly, technological unemployment. . . . The "price system" . . . was incompatible with technology. It irrationally vested power in commercial and financial interests whose sole concern was to maximize profits. The nontechnical values of businessmen and political decision makers made them incompetent to deal with technological society. The present crisis required the restructuring of society along more rational and scientific lines. In Scott's words, future society "must be adjusted to the energy producing values which can be regulated by scientific methods." This would organize society around the principle of production for use rather than for profit."[32]

This analysis seemed plausible at the height of the depression, and the committee's ideas found widespread support. Although the group officially eschewed politics, claiming instead that they were only advocating that the one best way to organize production and society be scientifically formulated, the movement that they engendered began to have political repercussions.

The committee was deliberately vague about specific plans for reorganizing society, preferring to focus on their research efforts, but the public, captivated by their analyses of the depression and promises of material abundance, became impatient to design and implement a "technocratic society." By late 1932, various groups across the United States were calling themselves "technocrats" and proposing various reforms. In Chicago, the Technocratic Party was formed, and another group began to form a ruling body of technical experts; in California, various groups advocated planning, more power for engineers, and a technocratic president.[33] As William Akin points out, one striking aspect of these diverse groups was their lack of concern about the authoritarian implications of their ideas; one group said that "the individual must subordinate himself to the community," which would be ruled by engineers; another group advocated taking the Soviet Union as the model for the new technocratic society.[34]

Increasingly pressured by society and the press to take a stand, the committee found it difficult to agree upon specific plans for social reorganization. Howard Scott set forth his own vision of a new distribution system derived from an energy system of value in 1933, but his scenario did not address the reorganization of production and did not find widespread support. Rautenstrauch set forth his own ideas, which centered on limiting the role of engineers, distancing himself from the more authoritarian proposals of a "rule by engineers," and stressing his commitment to democracy and representative government.[35] In 1933, the Committee on Technocracy disbanded, largely because of the divergence between Scott and Rautenstrauch as well as widespread criticism of Scott. Technocrats, both within the committee and in the country as a whole, found it easier to agree on criticisms of the status quo than on an alternative.

However, technocracy lived on as a decentralized and unorganized movement: Enthusiasm for technocracy remained high in dozens of cities in the Midwest, the Pacific Northwest, California, and western Canada until late in 1934. In the early stages it appealed to marginal groups, socialists, promoters, the economically hard hit lower middle class, and the elderly. As the movement grew its appeal reached urban Anglo-Saxon middle-class professionals and white-collar workers who were alienated from the major political parties.[36]

In New York, two separate groups formed: the Continental Committee on Technocracy, centered around Harold Loeb, a novelist; and Technocracy, Inc., centered around Scott. The Continental Committee, more concerned with national organizing, claimed 250,000 members in seventy locals by May 1933.[37] Technocracy, Inc., featured a uniform, consisting of a "well-tailored double-breasted suit, gray shirt, and blue necktie, with a monad insignia on the lapel";[38] its members even saluted Scott in public.

Loeb wrote *Life in a Technocracy* (1933) to consolidate and promote his ideas.[39] More of a humanistic intellectual than Scott or Rautenstrauch, Loeb aimed to popularize the ideas of technocracy by combining a simplified economic analysis, an equally simple technological determinism, and a utopian social projection of what a technocratic society would entail. The economics of the present crisis boiled down to this:

> Men live by the production, distribution, and consumption of goods. Goods are produced and distributed by effort. The incentive of effort is profit. Profit depends on price. Price depends on scarcity. Therefore, the life of man, under the capitalistic system, depends on the scarcity of goods.
> *And the scarcity of goods is being progressively destroyed by the application of science to production, known in its latest phase as technology.*[40] (emph. in original)

For Loeb, science and engineering come to the rescue, however, for "how best to produce and distribute the products of man's efforts is an engineering problem for each detail of which there is always a right answer."[41] Once society was organized in a more rational way, material abundance, individual freedom, and all humanistic values could be realized. Specific details of technocratic reorganization were left somewhat vague, although a national regulatory commission composed of representatives of each of ninety-two basic industries was envisioned to coordinate economic activities.

Loeb's technological determinism pointed toward a utopian teleology:

The irrestible succession continues. Again the nature of machinery is being altered. A continuous process is being substituted for the flexible batch process. Forced, routinary labor is needed less and less. Supervisory labor is no task for slaves. Gradually by abolishing scarcity and thus subverting the profit system dependent on scarcity, technological methods of production are compelling the adoption of a new idea system. . . . It is in this sense that technocracy is inevitable. Technological processes will compel a social system congenial to their operation or they will ruin the state.[42]

The "inevitable" technocracy is described in highly utopian terms, as freeing people from political tyranny and unhappy personal relationships, and as giving people the leisure to self-actualize. Even mating practices, once economic considerations are abolished, would be more eugenically sound:

A technocracy, then, should in time produce a race of man superior in quality to any now known on earth, a society more exciting, interesting, and variegated than has ever been possible, and a nation in which no individual should be unhappy or discontented for remediable causes.[43]

By the mid-1930s, the Technocracy movement was in decline. Technocracy, Inc., continues to exist even today, although its constituency is small.[44] Most historians have attributed the demise of the Technocracy movement to the rise of the New Deal, considered to be a more democratic method of accomplishing the planning and economic reconstruction that the Technocrats had called for.[45] The elitist and even fascistic overtones of the Technocracy movement undermined its popular appeal as a political movement. And yet, the technocratic vision was prescient in some ways:

The modern postindustrial state—with its centralization, its emphasis on replacing politics with administrative decisions, and its meritocratic elite of specially trained experts—bears a more striking resemblance to the progressive formulation, which was the starting point for the technocrats. The progressive intellectuals, progressive engineers, and scientific managers of the early twentieth century saw the outlines of the future political economy with amazing clarity. But the "immensely enriched and broadened life within reach of all," which Harlow Person predicted, remains a dream that technology and engineering rationality seem incapable of fulfilling.[46]

MANAGERIAL SOCIETY, THE NEW INDUSTRIAL STATE, AND POSTINDUSTRIALISM

James Burnham's *The Managerial Revolution* (1941) had certain affinities with the Technocrats' ideas but attempted to ground them in more sociological analyses of major changes in institutions, ideologies, and societal elites.[47] For Burnham, a social revolution was occurring that was as profound in its implications at the transition from feudalism to capitalism. Inspired by the New Deal and the Technocracy movement, and writing in the context of World War II, Burnham sought to more adequately theorize sociological change.

Like the Technocrats, Burnham stressed the increased centrality of technical expertise to the production process and the corresponding gap between this expertise and the level of knowledge of the ordinary worker. The technical obsolescence of ordinary workers implied that traditional Marxism is naive and outdated; the transition that was occurring was not from capitalism to socialism but from capitalism/socialism to managerial society. For Burnham, managers were on the ascendancy in capitalist and socialist societies alike, rendering such distinctions obsolete and creating a fundamentally different type of society.

Central to Burnham's analysis was the increased importance of the state, which he argued was growing in size and societal influence, and which he saw as "the 'property' of the managers,"[48] making managers a new ruling class, presiding over a large and powerful bureaucratic apparatus. For Burnham it was not engineers and technical experts who were assuming control, however important their function, for they did not have sufficient political power. Within the production process, as within society as a whole, managers were the dominant class.

> All the necessary workers, skilled and unskilled, and all the industrial scientists will not turn out automobiles. The diverse tasks must be organized, co-ordinated, so that the different materials, tools, machines, plants, workers are all available at the proper place and moment and in the proper numbers. This task of direction and co-ordination is itself a highly specialized function. . . . But it is a mistake (which was made by Veblen, among others) to confuse this direction and co-ordinating function with the scientific and engineering work.[49]

Like the Technocrats, Burnham wrote of the inevitable demise of capitalism. Certainly entrepreneurial capitalism, where owners manage their own firms (simple control), he saw as largely superseded. For Burnham, as for the Technocrats, contemporary management is too complex for most capitalists.

> Through changes in the technique of production, the functions of management become more distinctive, more complex, more specialized, and more crucial to the whole process of production, thus serving to set off those who perform those functions as a separate group or class in society; and at the same time those who formerly carried out what functions there were of management, the bourgeoisie, themselves withdraw from management.[50]

Burnham makes the further functional distinction between managers, executives (whose job is to ensure that the firm makes a profit), finance capitalists, and stockholders. He argues that not only are these functions typically assigned to different people, but that the managerial function alone is necessary to the process of production. In fact, he argues that finance capitalists and stockholders are typically quite removed from the actual production process, spending their time in leisure, travel, and other pursuits.

A decade earlier, Berle and Means had addressed similar issues, arguing that corporate ownership and corporate control were diverging, as corporate executives were no longer accountable to their stockholders, and envisioning that corporate executives should become professional administrators, arbitrating the claims of owners, employees, and consumers in a "purely neutral technocracy."[51] Despite some obvious similarities, Burnham differentiates his analysis from Berle and Means's separation of ownership and control. Berle and Means fail to distinguish between managers and executives and also minimize the close ties between capitalist executives and finance capitalists. Moreover, Burnham contends that, in the final analysis, "ownership means control,"[52] and if the two are in reality separated, then ownership has become rather meaningless.

Burnham admits that managers continue to have close ties to capitalists and that they are in some sense the servants of capitalist interests. However, his argument is that, given the centrality of managerial expertise to the production process, managers are quite capable of "turning on their masters."

> The instruments of production are the seat of social domination; who controls them in fact not in name, controls society, for they are the

means whereby society lives. The fact today is that the control of the big capitalists, the control based upon capitalist private property rights, over the instruments of production and their operation is, though still real, growing tenuous, indirect, intermittent. . . . Throughout industry, de facto control by the managers over the actual processes of production is rapidly growing in terms both of the aspects of production to which it extends and the times in which it is exercised . . . owners, in the legal and historical capitalist meaning, have scarcely anything to do with the corporations beyond drawing dividends when the managers grant them.

With capitalism collapsing and on its way out, the ruling capitalist class as a whole is being replaced by a new ruling class.[53]

Although Burnham was premature in proclaiming the imminent demise of capitalism, he nonetheless was prescient in analyzing certain aspects of its transformation. He recognized that in managerial society, technological developments and new types of technical and political crises would become increasingly salient.[54] Moreover, Burnham was not as naive as certain of the Technocrats in his evaluation of the managerial class: He saw it as an exploiting class and viewed the politics of managerial society as potentially fascistic and probably undemocratic, at least initially. However, according to his analysis, managerial society would contain the seeds of democracy, due to the fact that efficient planning necessitates some knowledge of public opinion.

The managerial economy cannot operate without a considerable degree of centralized planning. But in planning and co-ordinating the economic process, one of the factors that must be taken into account is the state of mind of the people, including something of their wants and of their reactions to the work they are doing. Unless these are known, at least roughly, even reasonable efficiency in production is difficult. But totalitarian dictatorship makes it very hard . . . to get any information on the actual state of mind of the people: no one is free to give unbiased information, and the ruling group becomes more and more liable to miscalculate, with the risk of having the social machine break down. A certain measure of democracy makes it easier for the ruling class to get more, and more accurate, information."[55]

However, Burnham also realized that the centralized economy of managerial society would make political opposition difficult and that democracy was far from inevitable.

Just as the Technocracy movement was partly inspired by the

depression, and Burnham's analysis of managerial society by the New Deal, the postwar optimism and prolonged economic prosperity gave rise to theories of the "new industrial state," "postindustrialism" and the "end of ideology."[56]

John Kenneth Galbraith's analysis of the new industrial state has affinities with both the technocrats and with Burnham. He attributes changes in the industrial system since World War II to several factors: the increased prevalence of sophisticated technology, the management of large corporations by managers rather than by entrepreneurs, and the increased size and transformed economic role of the state. Corollary changes include the growth of advertising, the decline of the trade union, and the expansion of higher education.[57] Galbraith sees all of these developments as part of a "matrix of change": Advanced technology implies capital investment, larger organizations, and increased need to plan; increased productivity implies affluence, discretionary income, and a greater capacity to manage demand through advertising and state control. The new industrial state is

> an economic system which, whatever its formal ideological billing, is in substantial part a planned economy. The initiative in deciding what is to be produced comes not from the sovereign consumer. . . . Rather it comes from the great producing organization which reaches forward to control the markets that it is presumed to serve and, beyond, to bend the customer to its needs. And, in so doing, it deeply influences his values and beliefs . . . there is a broad convergence between industrial systems. The imperatives of technology and organization, not the images of ideology, are what determine the shape of economic society.[58]

Central to the new industrial state is what Galbraith terms the "technostructure." Although Galbraith agrees with Burnham that entrepreneurial capitalism has largely been superseded by managerial society and that management effectively runs large corporations, he agrees with the Technocrats that managers are less central to the new industrial state than technical specialists:

> [Management] includes . . . only a small proportion of those who, as participants, contribute information to group decisions. This latter group is very large; it extends from the most senior officials of the corporation to where it meets, at the outer perimeter, the white and blue collar workers whose function is to conform more or less mechanically to instruction or routine. It embraces all who bring specialized knowledge, talent, or expe-

rience to group decision-making. This, not the management, is the guiding intelligence—the brain—of the enterprise. . . . I propose to call this organization the Technostructure.[59]

For Galbraith, as for the Technocrats, there is a tension, or contradiction, between the interests of the technostructure and capitalism: The technostructure relies on planning, rather than market imperatives, and does not emphasize profit maximization, as its members' salaries do not vary with profits. For members of the technostructure, as for the new industrial society more generally, new motivations and goals become salient: autonomy, technological virtuosity and innovation, the pursuit of "significant" work that addresses important social goals. In addition, secure earnings, price stability, and the management of demand all serve the technostructure's interest in planning.

The technostructure is heavily dependent upon what Galbraith terms the "educational and scientific estate," and vice versa, and yet the two are in conflict with one another in several ways. The intelligentsia tends to view business methods, particularly the cruder forms of advertising, with contempt. The technostructure emphasizes group decision making, whereas intellectuals, particularly humanistic and artistic ones, emphasize individual effort. Most importantly, however, institutions of higher education ideally serve to inspire critical reflection upon the societal status quo:

> The industrial system, by making trained and educated manpower the decisive factor of production, requires a highly developed educational system. If the educational system serves generally the beliefs of the industrial system, the influence and monolithic character of the latter will be enhanced. By the same token, should it be superior to and independent of the industrial system, it can be the necessary force for skepticism, emancipation, and pluralism.[60]

For Galbraith, intellectuals are less constrained politically than members of the technostructure and have the capacity to formulate and put into effect alternative social goals. The industrial system and the state are a powerful coalition, but it is the intelligentsia that will serve to make them responsive to broader social purposes.

An even stronger emphasis on knowledge and educated workers is found in the analyses of postindustrial society. The manifesto of postindustrialism, Daniel Bell's *The Coming of Post-Industrial Society* (1973), focuses upon the new centrality of knowledge, science, and technology.[61] For Bell, postindustrialism does not represent a challenge to capitalism, for they represent different dimensions of contemporary society: *postindustrialism* describes the sociotechnical dimension, and *capitalism* the socioeconomic. What is being described is a new stage of capitalist development.

> The concept "post-industrial" is counterposed to that of "pre-industrial" and "industrial." A pre-industrial sector is primarily *extractive*, its economy based on agriculture, mining, fishing, timber, and other resources such as natural gas or oil. An industrial sector is primarily *fabricating*, using energy and machine technology, for the manufacture of goods. A post-industrial sector is one of *processing* in which telecommunications and computers are strategic for the exchange of information and knowledge. . . .
> Broadly speaking, if industrial society is based on machine technology, post-industrial society is shaped by an intellectual technology. And if capital and labor are the major structural features of industrial society, information and knowledge are those of the post-industrial society. For this reason, the social organization of a post-industrial sector is vastly different from an industrial sector.[62]

For Bell, the central difference between industrial and postindustrial society derives from the difference between industrial production and knowledge production. He contends that knowledge is intrinsically *social* and that knowledge production cannot be analyzed using a labor theory of value.

> Post-industrial society is characterized not by a labor theory but by a knowledge theory of value. It is the codification of knowledge that becomes directive of innovation. Yet knowledge, even when it is sold, remains also with the producer. It is a "collective good" in that, once it has been created, it is by its character available to all.[63]

In postindustrial society there is therefore a new centrality of human communication, networking, and knowledge production. Work becomes a game between persons, rather than an activity centered around material production. Like Galbraith, Bell also sees the state as playing a more central role in knowledge production, largely because of the fact that it is a "collective good."

Bell argues that postindustrial societies tend to generate a "knowledge class" of technical and professional workers, which is growing in size and by the year 2000 will be the largest single occupational group.

This emphasis on technical skill and education as prerequisites for elite status implies a more meritocratic means of reward allocation, although property and political position continue to be important bases of power as well. After a lengthy discussion of meritocracy, Bell concludes that, while not perfectly egalitarian, the postindustrial impetus toward meritocracy is the best and most realistic hope for social justice.

Another important dimension of postindustrial society concerns the relationship between technical decision making and politics. Bell is critical of the radical technocratic view: that rational judgment has displaced politics altogether. However, he accepts the idea, also a central theme of *The End of Ideology*, that "technical decision-making . . . can be viewed as the diametric opposite of ideology: the one calculating and instrumental, the other emotional and expressive."[64] Indeed, the relationship between technical and political decision making is what is of critical importance for Bell. Although he argues for the politicization of technical decision making, by dichotomizing the two he undermines this possibility. Moreover, his assumption that knowledge is a collective good ignores existing monopolies of knowledge that also tend to operate against the politicization of technical decision making. Like Galbraith (and others, such as Touraine), Bell implicitly assumes that more progressive intellectuals, such as social scientists, should become more integral to technical decision making and planning.[65] As Gouldner points out: "The New Class believes that the world should be governed by those possessing superior competence, wisdom, and science—that is, themselves."[66]

More recently, Fred Block has attempted to update and reformulate postindustrial theory.[67] For Block, the key issue is that the concept of industrial society no longer allows us to make sense of our world, and that "'postindustrial society' is the historical period that begins when the concept of industrial society ceases to provide an adequate account of actual social developments."[68] Such developments as the growth of the service sector and decline of goods production, the transformation of the workplace that has accompanied computerization, and related changes such as the greater participation of women in the work force make postindustrialism an appealing new theoretical paradigm. Consistent with the postindustrial tradition, Block sees recent socioeconomic developments in a very positive light, arguing that they provide the preconditions for a more thorough democratization of society.[69]

Block's critique of mainstream economics and its inability to comprehend contemporary reality is persuasive, and his suggestions for the sorts of political change that are needed in order to realize the positive democratic potential of postindustrial society represent a considerable improvement upon postindustrial theory. However, because he fails to address the shortcomings of Bell's version of postindustrialism, his work has some similar limitations: an overly optimistic view of the effects of computerization on work organizations, an exaggerated emphasis on skill enhancement and the elimination of routine work, a vision that is focused narrowly on the United States and fails to theorize the importance of the worldwide capitalist system. The concept of industrial society is not sufficient to understand contemporary socioeconomic reality, but the concept of postindustrial society is sufficiently flawed to make it of limited utility as well.

NEW CLASS THEORISTS

In recent years, much theoretical effort has gone into conceptualizing the "new class" of educated workers. Despite general agreement that technical, professional, and scientific workers are in the ascendancy as an occupational group, specific analyses of the nature of this group and its impact upon the capitalist class structure have varied widely.

New-working-class theorists, inspired by the social movements of the 1960s, and particularly May 1968 in France, have argued that educated workers comprise a revolutionary class inherently antagonistic to capitalism, comparable in mission to the proletariat.[70] Their basic argument is that educated work involves norms of autonomy and creativity that are inherently contradictory to capitalist control:

> A conflict which is most often latent, but overt and severe in an increasing number of areas, begins to oppose the most qualified workers to the logic of monopoly capitalism. . . . The fundamental contradiction is that between the requirements and criteria of profitability set by monopoly capital . . . and the inherent requirements of an autonomous, creative activity which is an end in itself: an activity which measures the scientific and technical potential of an enterprise in scientific and technical terms. . . . Technicians, engineers, students, and researchers discover that they are wage earners like the others, paid for a piece of work which is "good" only to the degree that it is profitable in the short run. They discover that long range research, creative work on original problems, and the love of workmanship are incompatible with the criteria of capitalist profitability.[71]

Although the various new-working-class theorists define the group somewhat differently,[72] all see educated workers as coming into contradiction with capitalism, a particular manifestation of the general contradiction between the forces and the relations of production. Another aspect of the argument concerns the fact that educated workers are sufficiently integral to advanced capitalist production to be particularly well suited to challenge and transform capitalism.

The potential conflict between the imperatives of knowledge and the exigencies of institutionalized bureaucratic power has also been a concern of theorists of Eastern European society.[73] Mallet, for instance, posits that "the growing demand of a technically and culturally maturing working class [will] cause the rigid framework of bureaucratic planning to burst,"[74] ultimately resulting in a truly socialist society. Rudolph Bahro, in an even more original analysis, abandons the concept of class consciousness as a means of understanding Eastern European society, substituting for it the concept of *surplus consciousness*, or consciousness that is no longer needed for production of the necessities of life. For Bahro, such surplus consciousness is most prevalent among scientific workers, technicians, and engineers—the new working class. He sees this group as particularly prone to frustration of their creative capacities and likely to become the vanguard of the cultural revolution, as their surplus consciousness becomes translated into "emancipatory interests." However, Bahro also realizes the potential for surplus consciousness to be drained into *compensatory interests*, such as consumption of material goods and orientation toward status-enhancing symbols.

Diametrically opposed to new-working-class theory is professional-managerial-class (PMC) theory.[75] While PMC theorists agree that professionals, technical experts, and managers comprise an occupational group that has been on the ascendancy in recent decades, they view this group as a new class within capitalist society, separate from, and antagonistic to, *both* workers and capitalists. According to this argument, "the professional-managerial workers exist . . . only by virtue of the expropriation of the skills and culture once indigenous to the working class,"[76] and it is therefore objectively antagonistic to the working class. In fact, the PMC's most essential and general function is seen as the reproduction of capitalist culture and capitalist class relations.

However, from the late nineteenth century, when the PMC emerged as a class, to the present, the PMC has also embodied antagonism to the capitalist class. The Ehrenreichs point to battles over academic freedom and other struggles for occupational autonomy, as well as the resistance of capitalist employers to PMC leaders' vision of "a technocratic transformation of society in which all aspects of life would be 'rationalized' according to expert knowledge."[77] Despite a certain professional autonomy, the PMC is largely subordinate to the capitalist class, although not without its own sources of power. In effect, the PMC is in a contradictory class location within capitalism.

Gouldner's theory of the New Class of intellectuals and technical intelligentsia expounds on their contradictory class situation.[78] Gouldner goes beyond both new-working-class theory and PMC theory by postulating the New Class as a "flawed universal class": elitist and self-seeking, and yet also generally progressive in its political impact. For Gouldner, the New Class is internally divided; he distinguishes between the technical and the humanistic intelligentsia, with the former viewed as more integral to capitalist production and hence more politically conservative.[79] However, he also sees the New Class as united by a "culture of careful and critical discourse" and an emphasis on autonomy from capitalist imperatives.[80]

Gouldner also effectively goes beyond Bell's emphasis on the collective nature of knowledge in postindustrial society. For Gouldner, the New Class is a "cultural bourgeoisie who appropriates privately the advantages of an historically and collectively produced cultural capital."[81] Just as traditional capitalism rests on an ideology of capital as enhancing productivity (although the primary concern of capitalists is with generating income), New Class ideology rests on similar claims:

> The new ideology holds that productivity depends primarily on science and technology and that the society's problems are solvable on a techno-logical basis, and with the use of educationally acquired technical competence. While this ideology de-politicizes the public realm, and, in part, *because* it does this, it cannot be understood simply as legitimating the *status quo*, for the ideology of the autonomous technological process deligitimates all other social classes than the New Class. The use of science and technology as a legitimating ideology serves the New Class, lauding the functions it performs, the skills it possesses, the educational credentials it owns, and thereby strengthens the New Class's claims on incomes *within the status quo in which it finds itself.*[82]

For all of these limitations, however, Gouldner also finds the New Class to be the most progressive force in modern society. Compared with both capitalist and socialist bureaucrats, the New Class is less authoritarian, more likely to rely on careful and critical discourse to persuade others. It also embraces social goals that go beyond profitability: ecology, open communication, and internationalism. And yet the New Class embodies a new type of elitism:

> The paradox of the New Class is that it is both emancipatory and elitist. . . . The New Class bears a culture of critical and careful discourse which is an historically emancipatory rationality. The new discourse (CCD) is the grounding for a critique of established forms of domination and provides an escape from tradition, but it also bears the seeds of a new domination. . . . The culture of discourse of the New Class seeks to *control* everything, its topic and itself, believing that such domination is the only road to truth. . . . The New Class sets itself above others, holding that its speech is better than theirs; that the examined life (*their* examination) is better than the unexamined life. . . . Even as it subverts old inequities, the New Class silently inaugurates a new hierarchy of the knowing, the knowledgeable, the reflexive and the insightful. Those who talk well, it is held, excel those who talk poorly or not at all. It is now no longer enough simply to be good. Now, one has to explain it.[83]

One political issue that has not received enough attention concerns the relationship between the New Class of technical experts and more traditional elites, such as capitalist owners, corporate managers, and political leaders. There is some evidence that decision makers are finding alliances with technocrats quite appealing and useful, allowing them to respond to "governance crises" by depoliticization and improved management and planning.[84] As Fischer points out, such alliances among experts and more traditional elites raise serious problems for democratic societies.

> A close alliance between elites and technocrats holds out the potential for a mutually beneficial strategy. . . . Technocratic ideology . . . holds out the possibility of a new—and often seductive—form of elite politics. Grounded in technical competence of professional expertise, such a system not only shrouds critical decisions in what would appear to the the logic of technical imperatives, it also erects stringent barriers to popular participation. Only those with knowledge (or credentials) can hope to participate in deciding the sophisticated issues confronting postindustrial society.[85]

POSTSTRUCTURALISM AND POSTMODERNISM

In recent years, poststructuralist and postmodernist thinkers have continued to explore the social and political implications of the increasing centrality of experts and expert knowledge in the contemporary world. Like many of their predecessors, they see both positive and negative potential in recent developments.

Michel Foucault took as one of his main tasks the exploration of the complex interrelationships between knowledge and power. For Foucault, knowledge is an essential constitutive element of power in the modern world, leading to distinct strategies of power/knowledge that must be documented, analyzed, and understood.

> Mechanisms of power in general have never been much studied by history. History has studied those who held power—anecdotal histories of kings and generals; contrasted with this there has been the history of economic processes and infrastructures. Again, distinct from this, we have had histories of institutions, of what has been viewed as a superstructural level in relation to the economy. But power in its strategies, at once general and detailed, and its mechanisms, has never been studied. What has been studied even less is the relation between power and knowledge, the articulation of each on the other. It has been a tradition for humanism to assume that once someone gains power he ceases to know. Power makes men mad, and those who govern are blind. . . .
>
> Now I have been trying to make visible the constant articulation I think there is of power on knowledge and of knowledge on power. We should not be content to say that power has a need for such-and-such a discovery, such-and-such a form of knowledge, but we should add that the exercise of power itself creates and causes to emerge new objects of knowledge and accumulates new bodies of information. One can understand nothing about economic science if one does not know how power and economic power are exercised in everyday life. . . .
>
> It is not possible for power to be exercised without knowledge, it is impossible for knowledge not to engender power.[86]

For Foucault, the relationship between power and knowledge is not unique to the contemporary period, but it has become more pronounced in recent years. In his specific historical studies, Foucault documents the emergence of the *repressive hypothesis* and the rise of *biopower*. The repressive hypothesis is "the view that truth is intrinsically opposed to power and therefore inevitably plays a liberating role;" biopower is the modern type of decentralized and pervasive domination through organization and control of bodies, in the alleged interest of rationality and productivity.[87] This repressive hypothesis, which began to emerge as a political technology during the seventeenth century, posits increasing societal repression over time, a repression that can only be undone by speaking openly about sexual issues and bodily needs—the idea that the truth can set one free. For Foucault, this widespread belief that truth and domination are fundamentally opposed becomes profoundly ideological and one of the main ways in which the operation of power/knowledge is hidden and disguised.

Counterposed to the repressive hypothesis is Foucault's concept of biopower. During the same period that the repressive hypothesis was emerging as an ideology, biopower began to be consolidated as a new way of exercising power. Biopower involves two main components: a concern with social issues and the modern individual as the focus of political attention, and the emphasis on the human body as an object to be manipulated into docility and productivity.[88] As a concomitant of biopower, "technologies of discipline" arose in factories, schools, prisons, hospitals, and so on. Of course, the rise of biopower is coterminous with the rise of capitalism, but for Foucault the underlying reality of biopower is more fundamental than the more visible changes in the mode of production.

Another dimension of the emergence of biopower and its disciplinary technologies is the development of the social sciences as new intellectual disciplines, which serve to generate knowledge and perfect disciplinary technology. Although at one level the social sciences have always spoken the language of critique and reform (of capitalism, the *status quo*), on a deeper level they are seen by Foucault as deeply complicitous with the rise of biopower. By generating categories of knowledge that warp individual and social reality, they create the very social problems that they purport to address.

> The advance of bio-power is contemporary with the appearance and proliferation of the very categories of anomalies—the delinquent, the pervert, and so on—that technologies of power and knowledge were supposedly designed to eliminate. The spread of normalization operates through the creation of abnormalities which it then must treat and reform. By identifying the anomalies scientifically, the technologies of bio-power are in a perfect position to supervise and administer them.
>
> This effectively transforms into a technical problem—and thence into a field for expanding power—what might otherwise be construed as a failure of the whole system of operation. Political technologies advance by taking what is essentially a political problem, removing it from the realm of political discourse, and recasting it in the neutral language of science. Once this is accomplished the problems have become technical ones for specialists to debate. . . . When there was resistance, or failure to

achieve its stated aims, this was construed as further proof of the need to reinforce and extend the power of the experts. A technical matrix was established. By definition, there ought to be a way of solving any technical problem. Once this matrix was established, the spread of bio-power was assured, for there was nothing else to appeal to; any other standard could be shown to be abnormal or to present merely technical problems.[89]

Biopower became institutionalized through the development of specific disciplinary technologies in prisons, schools, mental institutions, and factories. Central to diverse disciplinary technologies is the emphasis on examination and surveillance of individuals: collecting information to further the growth of knowledge/power. Although Foucault documents and analyzes diverse disciplinary technologies, the most sustained and exemplary model he presents is that of the Panopticon: "the diagram of a mechanism of power reduced to its ideal form . . . a figure of political technology that may and must be detached from any specific use."[90]

Jeremy Bentham's Panopticon (1791) was a general institutional plan that consisted of a large courtyard with a tower in the center surrounded by buildings (of several levels) divided into cells. Each cell had a large observation window opening onto the courtyard, through which the inmate could be seen by the supervisor in the tower, but through which the inmate could not see the supervisor in the tower. Hence, this one-way observation creates a nexus of constant, anonymous power and surveillance, one that implicates the supervisor as well as the inmates. The inmates, since they never know when they are being observed, are forced to constantly monitor their behavior; the supervisors are administratively constrained and controlled in a somewhat different way. The entire technology functions so as to insure docility, productivity, and the generation of new knowledge/power.[91] Obviously, this is a model that symbolically has wide relevance; in fact, as we shall see in chapter 4, it has recently been used as a model of the contemporary computerized workplace.

By emphasizing the fact that power is exercised through diverse disciplinary technologies and derived from knowledge in general, Foucault points toward an extremely decentralized vision of the contemporary political landscape: "micropower" relations that are "rooted deep in the social nexus, not reconstituted 'above' society as a supplementary structure whose radical effacement one could perhaps dream of."[92] What must be analyzed are the "multiple forms of subjugation that have a place and function within the social organism,"[93] and, given the centrality of knowledge, differences in "know-how and competence"[94] are a crucial dimension of the micropolitics of contemporary society. Moreover, for Foucault power presupposes resistance and struggle. What remains to be understood are the new forms of micropolitics: strategies and struggles within diverse institutions that derive from the new types of power/knowledge and the "specific rationalities" being generated today.[95]

Postmodernism, although it is a more general and cultural phenomenon, has certain affinities with poststructuralism. Frederick Jameson views postmodernism as the cultural dominant of late capitalism.[96] As such, its more sociological manifestations have been concerned with the new types of knowledge and new class groupings characteristic of late capitalism.

Lyotard's *The Postmodern Condition* is the postmodernist text that most directly addresses issues related to technocracy.[97] Lyotard sets forth propositions concerning the impact of "computer hegemony" on knowledge itself: that a certain "exteriorization" of knowledge with respect to the "knower" occurs, that knowledge will become more and more commodified and will be produced in order to be sold, and that knowledge will lose its "use value."[98] For Lyotard, knowledge becomes an important force of production in late capitalism, with access to and control of information becoming important political variables, both at the level of nation states and within particular workplaces.

> Increasingly, the central question is becoming who will have access to the information these machines must have in storage to guarantee that the right decisions are made. Access to data is, and will continue to be, the prerogative of experts of all stripes. The ruling class is and will continue to be the class of decision makers.[99]

Another dimension of the transformation of what passes for knowledge in contemporary societies concerns the growing dominance of scientific knowledge and the demise of other ways of knowing, particularly what Lyotard calls narrative knowledge. As a result of the "blossoming of techniques and technologies since the Second World War,"[100] scientific knowledge and instrumental reason have displaced narrative as the most fundamental way of knowing. As compared with narrative, scientific knowledge involves several differences: in terms of language, an emphasis on denotation and an exclusion of other language games; a separation from the more general language games of society; monopolization of competence by scientists; a reliance on argumentation and proof to insure validity of statements; an emphasis on "diachronic temporality," or a cumulative development of scientific knowledge.[101]

Narrative knowledge, on the other hand, "does not give priority to the question of its own legitimation and . . . it certifies itself in the pragmatics of its own transmission without having recourse to argumentation and proof."[102] While narrative discourse accepts science as one of many varieties of narrative cultures, science does not reciprocate with similar tolerance:

> The scientist questions the validity of narrative statements and concludes that they are never subject to argumentation or proof. He classifies them as belonging to a different mentality: savage, primitive, underdeveloped, backward, alienated, composed of opinions, customs, authority, prejudice, ignorance, ideology. Narratives are fables, myths, legends, fit only for women and children."[103]

Another dimension of the growing emphasis on scientific knowledge is the corresponding emphasis on *performativity*. Technology follows the principle of optimal performance: maximizing output and minimizing input. A corresponding emphasis on technical efficiency pervades contemporary society, to the extent that the performativity principle becomes more important than truth itself.

> The question (overt or implied) now asked by the professionalist student, the State, or institutions of higher education is no longer "is it true?" but "What use is it?" In the context of the mercantilization of knowledge, more often than not this question is equivalent to: "Is it saleable?" And in the context of power-growth: "Is it efficient?" . . . What no longer makes the grade is competence as defined by other criteria: true/false, just/unjust, etc.—and of course, low performativity in general.
> This creates the prospect for a vast market for competence in operational skills. Those who possess this kind of knowledge will be the object of offers or even seduction strategies. . . . What is of utmost importance is the capacity to actualize the relevant data for solving a problem "here and now," and to organize that data into an efficient strategy.[104]

Other corollaries of the dominance of performativity that Lyotard discusses are the concern with interdisciplinary studies and the reliance on teamwork to accomplish specific goals.

The emphasis on performativity also assumes a stable system, one in which input and output can be calculated so as to achieve a certain input/output ratio. However, as Lyotard points out, the idea of perfect control over a system also contradicts performativity by lowering performance levels. This dynamic can be seen most clearly in the way in which bureaucracies "stifle the systems or subsystems they control and asphyxiate themselves in the process."[105]

Finally, Lyotard sees computerization as capable of promoting either a highly controlled market system dominated by the performance criterion and the use of terror, or a more democratic society that would have its decision making enhanced by computerized data banks. In order for the latter scenario to occur, however, the public would have to have "free access to the memory and data banks."[106] Another aspect of Lyotard's vision of a more liberated society is the development of a "postmodern science," one that would focus on

> undecidables, the limits of precise control, conflicts characterized by incomplete information . . . catastrophes, and pragmatic paradoxes . . . it is producing not the known but the unknown. And it suggests a model of legitimation that has nothing to do with maximized performance, but has as its basis difference understood as paralogy.[107]

CONCLUSION

As we have seen, social theorists have analyzed the social role of experts and expert knowledge for at least two centuries. Not surprisingly, over time the prevailing societal perspective has shifted dramatically. What is most apparent is a steady growth of skepticism and pessimism.

In retrospect, Saint-Simon, Taylor, and Veblen seem naive in their credulous faith in the enlightened rule and universalist interests of scientific experts and the capacity for social emancipation through scientific engineering. By the time of Burnham, an incipient pessimism emerges concerning the authoritarian potential of the managerial class. Postindustrial scenarios express more optimism concerning the overall societal effects of the new centrality of knowledge, although not without some misgivings. New-working-class and professional-managerial-class theorists have taken diametrically opposed stances concerning the political role of technical experts in advanced capitalist society, a conflicted stance that is also exemplified by Gouldner's conclusion that the New Class is both emancipatory and elitist.

With the contemporary work of Foucault and Lyotard, the skepticism becomes deeper. Foucault has documented and analyzed the growing authoritarianism of knowledge itself, as well as the various ideologies that support it. For Lyotard, the hegemony of scientific, computerized knowledge and the performativity principle that it promotes have become so dangerous to democracy and human society that a fundamentally new conception of science and rationality is needed.

Chapter 3

THE TRANSFORMATION OF TECHNICAL CONTROL

In a curious irony the factory, first spawned by the industrial revolution—a revolution of new mass production technology—is now being strangled by the very same kind of . . . mass production that made it so uniquely productive when it burst on the scene 200 years ago. Today our mass production is a strategic millstone, dangerously inflexible in its ability to change products and output levels, while offering distasteful, monotonous jobs for a high proportion of employees.
—Skinner, "Wanted: Managers for the Factory of the Future"

[With advanced technology] real wealth manifests itself . . . and large industry reveals this—in the monstrous disproportion between the labour time applied, and its product, as well as in the qualitative imbalance between labour, reduced to a pure abstraction, and the power of the production process it superintends. Labour no longer appears so much to be included within the production process; rather, the human being comes to relate more as watchman and regulator to the production process itself. . . . He steps to the side of the production process instead of being its chief actor. In this transformation, it is neither the direct human labour he himself performs, nor the time during which he works, but rather the appropriation of his own general productive power, his understanding of nature and his mastery over it by virtue of his presence as a social body—it is, in a word, the development of the social individual which appears as the great foundation-stone of production and of wealth.
—Karl Marx, Grundrisse

As we saw in chapter 1, the post-World War II period has witnessed an an "unprecedented restructuring of the organization of production".[1] However, conceptualizations of these changes in production workplaces have varied widely. Braverman and those contemporary social scientists influenced by his perspective have tended to view advanced technology as promoting a more stringent and all-encompassing form of managerial control over production work forces, with further erosion of worker skill and judgment.[2] At the other extreme, the sociotechnical view of computerized workplaces stresses the potential of computerized technology to be used to enhance worker control over the work process and promote the integration of learning and work.[3]

Advanced technology is neither monolithic nor deterministic, and it is therefore not surprising that diverse findings and interpretations of its effects are apparent. In order to make sense of these divergent views, a closer examination of both these varied conceptualizations and of some of the recent empirical literature on the transformation of production workplaces is needed. Only through concrete analyses of changes at the point of production can we understand the interrelationships among social, political, and technological changes and the ways in which these interrelationships have resulted in transformations of organizational structure.

RECENT CONCEPTUALIZATIONS OF CHANGES IN TECHNICAL CONTROL

Although production technology has been used to enhance managerial control at least since the beginning of capitalism,[4] it was only in the early years of the twentieth century, with the development of assembly-line technology and scientific management, that technical control began to achieve full expression. Harry Braverman, in a well-known analysis, studied the ways in which diverse technologies, when used in a capitalist/Taylorist context, have served to enhance managerial control of the labor process by divorcing conception from execution, deskilling work, and restricting worker autonomy and responsibility.[5]

Those contemporary analysts of the labor process working within a Bravermanian framework have provided considerable empirical documentation that Taylorist intentions and results are still apparent in workplaces utilizing more advanced technology.[6] According to these analysts, advanced computer and telecommunications technology has served to either eliminate or deskill jobs, and has also increased the capacity of management to monitor workers through the technology. They contend that the gap between conception and execution has widened, resulting in a more extreme polarization among workers. The neo-Bravermanians therefore contend that whatever the positive potential of advanced technology, in practice this potential has been thwarted and subsumed by prevailing capitalist social relations so as to extend and perfect technical control.[7]

However, divergent analyses of the impact of advanced technology on workplaces have also been apparent. In an early study of automation, Blauner studied four different types of technology in four different industries: craft technology (printing), machine-tending technology (textiles), assembly-line technology (automobiles), and continuous-process production or automation (chemicals).[8] Blauner found that, rather than deskilling, the automated chemical plants he studied exhibited an upskilling trend, due to enhanced understanding of the total operations of the plant on the part of individual operators and increased emphasis on maintenance skills. He also found that the operators of the continuous-process technology were less stringently supervised, were less isolated because they worked in teams, had more opportunities to learn and grow on the job, and exhibited less alienation in their work than workers in industries with less advanced technology.[9] Blauner concludes that "alienation has traveled a course that could be charted on a graph by means of an inverted U-curve":[10] alienation is low in craft work and among workers operating continuous-process technology, and highest among machine tenders and assembly-line workers. He finds level of skill following a similar trajectory.

More recently, similarly optimistic depictions of the potential benefits of advanced computer automation have come from sociotechnical analysts.[11] Based on the sociotechnical conception of work design, first developed by Eric Trist and others at the Tavistock Institute in London in the post-World War II period, this perspective argues that work organizations are complex systems of both social and technical relations, and that social and technological aspects of the organization must be designed to complement one another.[12] The same technology may be implemented in the context of very different social relations, yielding work organizations with divergent effects and variable degrees of success.[13]

The sociotechnical outlook is consistent with postindustrial theory in that advanced technology, particularly integrated computerized production systems, is seen as capable of being designed and implemented so as to expand worker skill and control. Sociotechnical analysts argue that more advanced technology, which is both very expensive and inherently vulnerable to technical problems, leads to a heightened dependence on remaining operators in order to insure productivity, cost-effectiveness, and quality control. Given the fact that automation implies a tendency to reduce worker input to monitoring of the equipment, social restructuring is necessary to avoid worker alienation and boredom and to realize technological potential.

The sociotechnical solution is to advocate teams of workers, job rotation, worker learning about the total operation of the plant, and worker monitoring, not only of particular tasks but of the entire operation of the plant. The technical flexibility of the integrated computerized plant (parts can be put through variable sequences of machinery, and design can be varied by differential computer programming both of the sequence and of particular operations) is therefore complemented by social flexibility: decentralization and teams of multiskilled workers.

Decentralized control, using microcomputers on the shop floor, is encouraged as an alternative to the highly centralized, fully integrated model.[14] Whereas centralized, highly integrated systems have Taylorist implications for workers, the emphasis on more decentralized systems "creates a conception of work in which the worker's capacity to learn, to adapt, and to regulate the evolving controls becomes central to the machine system's developmental potential."[15] Sociotechnical analysts see the tension between worker learning and managerial control as fundamental and argue that, given the inherent tendency for technical failure, expanded worker learning, responsibility, and control are prerequisites for the successful operation of cybernetic production systems.

The sociotechnical view sees workers in computerized plants as losing execution functions but as potentially gaining in knowledge and power, as their judgment becomes increasingly important. The sophisticated technology is thus used to expand and enhance the work situation: "As knowledge is incorporated into machines, workers can reinvolve themselves at a wider and more comprehensive level of production."[16] Social design therefore becomes important in creating a work environment that will promote worker learning and the integration of work with planning: teams of multiskilled workers paid generous salaries (rather than wages or by piece rates), job rotation, and an emphasis on technical coordination rather than traditional control.

Workers get rewarded for quality rather than quantity, and in some cases for learning rather than job performance.

A recent analysis of computerization that is consistent with the sociotechnical perspective is Shoshana Zuboff's *In the Age of the Smart Machine*. For Zuboff, an important dimension of computerized technology that is often overlooked is its capacity to *informate* jobs:

> Information technology is characterized by a fundamental duality that has not yet been fully appreciated. On the one hand, the technology can be applied to automating operations according to a logic that hardly differs from that of the nineteenth-century machine system—replace the human body with a technology that enables the same processes to be performed with more continuity and control. On the other, the same technology simultaneously generates information about the underlying productive and administrative processes through which an organization accomplishes its work. It provides a deeper level of transparency to activities that had been either partially or completely opaque. In this way information technology supersedes the traditional logic of automation. The word that I have coined to describe this unique capacity is *informate*. Activities, events, and objects are translated into and made visible by information when a technology *informates* as well as *automates*.[17]

Zuboff's detailed case studies of both "manufacturing environments" and "office environments" document how this new informating capacity, whether developed or thwarted by managerial purposes and organizational design, is transforming work environments. When the automating potential is emphasized, expert-sector workers embed knowledge in the technological system to maximize efficiency, with nonexpert-sector workers being deskilled and deemphasized. Conversely, with an informating strategy, workers at all levels are given the opportunity to acquire and utilize the rich data that computer technology affords. They are encouraged and trained to gain comprehensive understanding: not only *how* to do their jobs but also *why* the system functions as it does. With an informating strategy, organizations are less bifurcated, as nonexpert workers become paraprofessionals, interacting intensively with the technology, one another, and management.

Highly divergent analyses, then, have shaped the debate about the impact of advanced technology on the workplace. In order to more fully understand and evaluate these perspectives, we turn to a more detailed examination of some of the recent empirical literature.

NUMERICAL CONTROL

Computer automation has its roots in the technological changes that began during World War II. David Noble, for instance, has analyzed the emergence of the "military-industrial-scientific complex" that became dominant during and after World War II, contending that it set the stage for the increased automation of production and shaped technological development in important ways.[18] He shows how both the war itself, with its urgent production quotas and labor upheavals, and the ensuing cold war promoted a type of technological rationalization that was to have a lasting impact on U.S. society and the world.

Noble argues: "The advantages of automation were heralded by people with a range of motives—control over the work force, technical enthusiasm for fascinating new devices, an ideological faith in mechanization as the embodiment of progress, a genuine interest in producing more goods more cheaply, concern about meeting military objectives."[19] Moreover, a perceived shortage of skilled workers fueled the enthusiasm for automation, creating a self-fulfilling prophecy.[20]

Noble contends that militarist and capitalist imperatives toward centralized control shaped the form of capital-intensive automation far more than did economic or technical concerns. He uses as his main illustration numerical control of machine tool production, which by the mid-1950s "had emerged from the military drawing boards . . . it became the unique and ubiquitous answer to the manufacturing challenge of programmable automation, not only in the United States but throughout the industrialized world."[21] This emergence of numerical control is worth examining in more detail.

Numerical control (N/C) involves the transfer of machinist skill to a computer program, which can then be use to automatically operate the machine.[22] The programmer analyzes the part to be made and the machinist movements needed to produce the part, then writes a program of instructions to the machine designed to mimic the machinist's role. The machinist thus becomes a monitor of the automatic production process, leading to allegations of deskilling. Machining skill becomes "company-owned" and controlled, a Taylorist motivation that is clear from the relevant managerial and engineering literature.[23]

An alternative to N/C, which emerged about the same time, was record-playback (R/P) technology. Noble describes the differences between the two in the following way:

R/P programming resembled the approach used with the later player pianos. Here, machinist skill was viewed more like music-making . . . and was acknowledged to be the fundamental and irreplaceable store of the inherited intelligence of metalworking production. Hence, the purpose of R/P was not to eliminate that skill altogether through the use of some formal substitute, but rather to reproduce it as faithfully as possible in order to multiply, magnify, or extend its range.[24]

R/P thus differs from N/C primarily in terms of who does the programming: workers or professional programmers. At the time when both emerged, R/P was probably somewhat technically superior, producing a program that more carefully emulated the subtle machining operations of the machinists. Yet N/C was developed, and R/P was not. Why?

In contrast to conceptualizations of technological development that stress Darwinian natural selection based on objective criteria of technical superiority, Noble's argument is that R/P techniques were not chosen due to political reasons (although technical reasons tended to be given). Although both techniques were designed to mechanize the labor process and reduce the need for skilled workers, N/C circumvented worker skill more completely by utilizing programmers. R/P relied on the recording of skilled worker performance, resulting in technically comparable results, but with an ongoing reliance on at least some skilled workers. N/C thus embodied and promoted more complete "expert" control over the labor process, whereas R/P would have implied greater worker input and control over programming.[25] R/P techniques, like other "programming by doing" approaches, while technically feasible, have typically not been adopted.[26]

Although developed during the 1950s, numerically controlled machines were not widely introduced until the 1970s, due to both technical and political difficulties. With the embodiment of worker skill in the program, the machinist on a N/C machine becomes a monitor of production, a tender of the machine or of several machines—a "machining center." Shaiken found that perceived deskilling of the job has led to job dissatisfaction and feelings of underutilization among skilled machinists, mitigated by feelings of "You can't stop progress."[27]

Harley Shaiken also observed conflicts between programmers and machinists (professional/worker conflicts) over the relative advantages of concrete skill versus abstract knowledge, as well as the locus of power. Management has tended to ally with the programmers in an expert coalition:

> Top management . . . was siding with the programmers in order to pre-
> serve its authority on the shop floor, realizing that the ability of machin-
> ists to produce more also might mean the ability to produce less. Pro-
> ductivity was sacrificed in the short run for the chance of predictable
> output in the long run.[28]

Conflicts between programmers and managers have also arisen,
derived from differences in their respective sources of authority. Kraft,
for instance, found conflicts arising between the "analyst as expert"
and the "analyst as manager," conflicts that are exacerbated when managers have limited technical knowledge but must attempt to retain final
authority over technical decisions.[29] Hirschhorn found that managers of
utility companies were often threatened by engineering expertise and
perceived threats to managerial power; one manager said, "'We don't
want an engineer who would be too inquisitive, who would go around
analyzing designs instead of operating the system.'"[30] In general, however, conflicts between technical experts and managers have been less
frequent than those between workers and either managers or experts.

There are both technical advantages and important technical limitations to N/C technology. N/C is both sophisticated and flexible,
making possible the production of any part that can be designed. However, as conception is increasingly severed from the shop floor, the
potential for error is increased, and the complex technology becomes
more vulnerable to failure.[31] Most programmers have had little or no
practical experience with manual machining, and if the work force is
deskilled, this practical knowledge may be increasingly scarce, making it difficult to adequately respond to system failures or errors.
Worker judgment and skill are still needed, as even the most sophisticated N/C machines cannot compensate for subtle worker skill in
adjusting technique to shifting, and only partially predictable, conditions. As one manager recognized, "part programming is not strictly a
science but more of an art or a craft, and it is very difficult to automate
a craft."[32] There is not a "one best way" to make a part.

Despite ongoing technical problems, N/C has become more
widespread in recent years. The garment industry, for instance, has
been revolutionized by computerization. As early as 1974, Braverman
wrote:

> Advanced production methods are copied from sheetmetal and boiler-
> shop techniques: die-cutting to replace hand cutting, pattern-grading
> equipment which produces different sizes of a master pattern, etc. There

is a photoline tracer which guides a sewing head along the path of a pattern placed in a control unit. Improving on this, a photoelectric control is used to guide a sewing head along the edge of the fabric. In these latter innovations we see the manner in which science and technology apply similar principles to dissimilar processes, since the same control principles may be applied to complex contours, whether on steel or cloth.[33]

An intensive case study of recent changes in the garment industry is provided by Cynthia Cockburn, who studied the impact of computerization in two garment factories in England.[34] Computer-aided design (CAD) and computer-aided manufacture (CAM) have transformed the making of patterns, as well as the cutting of pattern pieces and the sewing of garments. In particular, Cockburn argues that the skilled manual process of pattern making and cutting has been deskilled, fragmented, and feminized with computerization. A master garment pattern, called up on the computer screen, can be modified (shortened, lengthened, pleats added, etc.) on the screen and reproduced in paper or cloth, in various sizes, through CAM. Computerized numerical control of sewing machines also rationalizes the making of garments, although, as Cockburn points out, "engineers have found it impossible to invent a machine that can handle soft and floppy fabric without the intervention of a skilled hand."[35]

Computerization of the garment industry has enhanced managerial control: "It puts management in the driving seat," as one senior manager told Cockburn.[36] As knowledge and skill are stored in computer memory, there is less reliance on skilled workers and more ability to monitor the production process, individual workers, and utilization of materials. Skilled craft workers have been replaced by a polarized and sex-segregated workforce, with skilled male technicians and semiskilled female operatives:

> What has happened in the process of technological change in the pattern and cutting rooms . . . is that men that once produced their consumer goods by means of simple tools (scissors, pencils) entirely under their own control, have been turned into operators (or, more commonly, replaced by women operators) working on machines that are under the technological sway and authority of men with technical skills.[37]

Management's expansion of technical control through N/C is not without its limitations and contradictions, however. In its pursuit of more complete control, management has been limited for political as well as technical reasons:

The cost-effectiveness of N/C, then, was dependent upon optimum utilization of the equipment, and this could only occur with effective maintenance of the machinery, careful coordination of the production process as a whole, and efficient machine operation. All of these factors, however, were dependent, in the final analysis, not only upon greater management supervision, planning, or use of computers, but upon the initiative, skill, judgment, and cooperation of the work force. Here, then, lay the central contradiction of N/C use: in its effort to extend its control over production, management set out to deskill, discipline, and displace the very people upon whose knowledge and goodwill the optimum utilization, and thus cost-effectiveness, of N/C ultimately depended."[38]

Technical considerations, then, are closely intertwined with political and social dimensions. As Shaiken points out, "decisions about how to program the machine tool are choices about how to organize the workplace."[39] One aspect of this political reorganization of the workplace is the increased importance of programmers as a new professional occupation within automated factories. In effect, management has replaced a reliance on skilled craft workers with a reliance upon a new type of skilled worker: the professional programmer. Craft skill is deemphasized in favor of more formalized knowledge and education: "The most important objectives in the implementation of computer-aided manufacture are to convert the 'know-how' of manufacturing from an 'experience-based' technology to a 'science-based' technology . . . so that computers can be used to implement this 'know-how' in product design, in manufacturing planning, and for control on the shopfloor."[40] This change is having important political and organization ramifications in contemporary factories.

Issues related to programmer (and other computer professional) training—its content and who obtains it—are important variables in the three-way struggle among management, professionals, and workers over definitions of skill and authority.[41] Programming relies on both practical, concrete knowledge (e.g., of machine tooling) and on some more formalized engineering and mathematical knowledge. In some instances, machinists have received supplementary technical training and become programmers. However, if a central managerial motivation for automation is to transfer the locus of skill and power away from the shop floor, then management is unlikely to favor training machinists to become N/C programmers. Conflict between workers and management over the acquisition of technical training and jurisdiction over technical skill has occurred, with managers typically preferring to train managers (some of whom were formerly machinists) rather than machinists.[42]

With the development of minicomputers and more user-friendly computerized numerical control (CNC), new opportunities for the decentralization of programming have emerged. However, the technology can also be used to enhance surveillance, since both the time spent doing a given job and productivity can be monitored and compared with managerial guidelines. In practice, the technology has taken different political form in different plants, with decisions concerning the degree of worker input made by management. In some instances, skilled workers do the bulk of the programming and editing at their machines; in others, workers are allowed limited editing functions to minimize boredom; in others, the technology is used solely for surveillance of the workers who monitor its centralized operations.[43]

Both the design and the implementation of CNC technology are flexible enough to permit various uses. The contradiction between managers' need to control workers and their reliance on worker skill, judgment, and motivation is assessed, and addressed differently by different firms. However, in 1984 the Office of Technology Assessment concluded from its survey of metalworking shops that the dominant trend is toward "reorganizing production in such a way as to centralize control and reduce the overall skill requirements of the shop."[44] Noble contends that the authoritarian model is clearly dominant, even when sacrifices of productivity must be made, due to the overriding managerial orientation toward control.

PULP AND PAPER MILLS

Zuboff, who analyzed the process of computer automation in nine different work organizations, concludes that the most fundamental way in which working with computerized systems differs from noncomputerized work is that it is more *abstract*.

> The distinction in feedback is what separates the linotype operator from the clerical worker who inputs cold type, the engineer who works with computer-aided design from one who directly handles materials, the continuous process operator who reads information from a visual display unit from one who actually checks vat levels, and even the bill collector who works with an on-line . . . system from a predecessor who handled accounts cards. Computer-mediated work is the electronic manipulation of symbols. Instead of a sensual activity, it is an abstract one.[45]

In her industrial work sites (two paper mills and one pulp and paper mill), all recently computerized, Zuboff documents this shift from concrete to abstract work, as well as its implications for workers and managers. She found that workers highlighted four aspects of non-computerized work: *sentience*, or the fact that information was based on physical cues; *action dependence*, or the fact that skill is developed through physical performance; *context dependence*, the fact that skill only has meaning within a certain context; and *personalism*, or the felt linkage between the individual and the knowledge and skill he or she demonstrates.[46]

With computerization, the work process is transformed. The physical relationship between a worker and the work is lost, leading workers to say such things as "I miss being able to see it. You can see when the pulp runs over a vat. You know what's happening";[47] "We have to control our operations blind";[48] and "The difficulty is not being able to touch things."[49] Instead of a reliance on physical cues, these workers had been forced to adapt to monitoring the plant's operations via a computer terminal, but this adaptation had generally been problematic for workers, who mistrusted the computer's messages and gave examples of how an overreliance on computerized messages had resulted in workers' overlooking plant malfunctions. According to Zuboff, the workers tended to experience "epistemological distress."

For Zuboff, the more abstract work process characteristic of computer-mediated work points toward a new type of requisite skill: *intellective skill*. One paper mill manager described the new skill demands in the following way:

> The workers have an intuitive feel of what the process needs to be. . . . All of their senses are supplying data. But once they are in the control room, all they have to do is look at the screen. Things are concentrated right in front of you. You don't have sensory feedback. You have to draw inferences by watching the data, so you must understand the theory behind it. In the long run, you would like people who can take data and draw broad conclusions from it. They must be more scientific.[50]

Intellective skill, then, involves inferential reasoning from abstract cues, which in turn rests on a more comprehensive understanding of the logic of the computerized system.

Zuboff found that the workers in the paper mills had a difficult time adjusting to the shifting skill requirements. Their sense of competence was confused; one operator put it: "We never got paid to have ideas . . . we got paid to work."[51] Some workers felt panicky at the thought of being judged according to new norms of competence that they felt uncertain of meeting. Others felt that to think in this way was a managerial prerogative, dangerous to usurp. As one manager put it, "currently, managers make all the decisions. . . . Operators don't want to hear about alternatives. They have been trained to do, not to think. There is a fear of being punished if you think. This translates into a fear of the new technology."[52]

Managers also had difficulty adjusting to the new skills the workers were acquiring, fearing threats to managerial authority. One manager put it: "Managers perceive workers who have information as a threat. They are afraid of not being the 'expert'";[53] another said, "We have not given the operators the skills they need to exercise this kind of judgment because we don't trust them."[54] A worker showed awareness of these managerial attitudes when he said, "It seems management is afraid to let us learn too much about how this system operates. The more we know, the more we could sabotage it."[55] Clearly, computerization raises issues that go far beyond technical concerns, penetrating to the heart of authority and control arrangements in contemporary workplaces.

In general, Zuboff found a deep polarization between managers and workers, even as the technology created the potential for workers to take on more managerial tasks due to its informating capacity. Managers generally chose to bypass this informating capacity and maintain distinctions based on differences in education, access to information, and social class. One worker described the subtle differences in this way:

> Managers are in a hell of a position. They have to appear to be in control all the time. They all have college degrees, and they start talking all kinds of terminology. They get an idea and then tell you to do it.
>
> Managers are supposed to be superior. They have all the outward trappings, like an office, no time clock, no shift work, and they can be transferred. Being an operator doesn't have prestige because you can get dirty. When you've gone to college, you don't wear jeans . . .
>
> Management is in a yacht, and we are in a rowboat. We're raised different. Education is a big thing—we didn't go to a formal college and get degrees. We've got different personalities and interests. We like to fish and drink beer, and they . . . I'm not sure what they do.[56]

Zuboff also found that managers deliberately monopolized information and access to computerized data in order to maintain distinctions of rank. One worker, for instance, contended that managers need to believe that hourly people are dumb. They come out of school and have to show that they are smarter than the people who have been working here for so long. They can't share information with us, because only by holding onto knowledge that we don't have can they maintain their superiority and their ability to order us around.[57]

Zuboff found that the planning that was done in conjunction with computerization emphasized centralized control and that worker training was not included as a budget item in the $200 million conversion plan at one pulp mill. In addition to maintaining managerial prerogatives, she found that the systems designers contributed to the polarization by creating computerized systems that were deliberately "user unfriendly" for both managers and workers, but particularly for workers.

As the automating strategy is emphasized and workers become mere appendages of the technology, a self-fulfilling prophecy is created. If workers are not given training and the opportunity to develop a comprehensive understanding of the system, they lose the ability to effectively monitor the working of the system, and more automatic control and managerial supervision is necessary. Like Noble, Zuboff argues that this is counterproductive in the long run, but that managers are often willing to forego potential gains in productivity and worker satisfaction in order to maintain traditional power distinctions.

The polarization into expert and nonexpert sectors (with some divisions within the expert sector) that Zuboff found is also exacerbated by a growing ambiguity as to the role of middle management and supervisors. With both automating and informating strategies, the role of supervisors is reduced, if not eliminated. With an informating approach, workers take on supervisory functions, and with an automating strategy workers are supervised by the technology itself.[58] Zuboff's findings in the pulp and paper mills she studied indicate that a polarized workplace utilizing an automating strategy is generally the result of computerization.

THE NEWSPAPER INDUSTRY

In 1964 Blauner analyzed the printing trade as exemplary of craft work and technology.[59] However, the newspaper industry has been dramatically transformed by computerized technology, leading Arne Kalleberg et al. to conclude: "Surely, there are few other industries that, in the span of a single generation, so embody the remembrances of our preindustrial past and prefigure the future of an automated society."[60]

There are four distinct aspects of traditional newspaper production: (1)composition, or the composing of copy into type and pages; (2)platemaking, in which plates of the pages are made; (3)presswork, where the print is transferred to paper; and (4)the mailroom, where the finished papers are prepared for distribution. Each of these types of work has been fundamentally altered during the past few decades.

During the 1950s and 1960s, the work of printing compositors underwent a dramatic technological transformation, a change from hot-metal Linotype typesetting and manual page composition to computerized photocomposition. This change, in turn, led to an equally dramatic restructuring of class and gender relations in the workplace. Cynthia Cockburn studied compositors in Great Britain who had undergone this transformation, and the impact of the new technology on their work and their lives.[61] Their experiences are worth examining in some detail.

From the 1890s (when typesetting was mechanized) until the 1950s, the newspaper compositors' trade remained virtually unchanged,[62] centered around the use of Linotype technology for typesetting (with some hand typesetting) and manual composition of pages. This technological longevity contributed to a highly integrated craft organization and craft culture, almost exclusively male.[63]

Linotype operators use a keyboard similar to that of a typewriter, but considerably larger, with ninety keys (for upper and lower case, different typefaces, digits, other symbols) and with more space between letters. The operator begins with a rough draft of copy from the journalist. As a letter or symbol is pushed, a hollow mold of the letter is manually released into a chute, and the letter molds collect in an assembler. The operator must decide when to end the line, how to hyphenate words, and how to space the words and letters to justify the right margin. Once a line is finished, a lever is pushed that forces molten metal into the molds, resulting in a "line o' type" that is transferred to a waiting galley tray. The molds are then reused, as the galley accumulates lines. Eventually the galleys are inked up and used to produce a page proof, which is proofread. The paste-up compositor then takes the corrected galleys and lays them out into pages, along with illustrations, advertisements, and so on.

Photocomposition, or "cold composition" as it sometimes called to differentiate it from "hot-metal" Linotype work, differs fundamentally. In a process similar to word processing, the operator sits at a computerized VDT. A typewriter keyboard is used, which not only has far fewer keys but also has a completely different arrangement of keys in closer proximity to one another. The process of ending lines, dividing words, and justifying margins is computerized. On-line systems link journalists, editors, advertising personnel, proofreaders, managers, and (in some cases) the compositors. In some cases, "direct entry" of copy is used, bypassing the photocompositor. Entire newspaper pages can be made up on the computer screen, eliminating manual composition of pages. After photocomposition (by whomever), the copy is printed using offset lithography.

By 1955, cold composition was being developed in the United States, and by 1974 four-fifths of U.S. dailies used this method; the number of compositors declined dramatically, with a 52% reduction (from 14,500 to 6900) between 1970 and 1983 and further reductions projected.[64] Britain did not embrace the technology so readily, but by the 1970s and 1980s, the majority of local newspapers had shifted to cold composition.

This dramatic technological shift has had equally dramatic social and political implications for the newspaper trade. Aside from the reduced numbers of compositors and the possibility that their job will increasingly be phased out in the trend toward direct input by the originator of the copy, remaining compositors have undergone a radical transformation of working conditions, labor process, and work culture.

Working conditions with the old hot-metal technology were dirty, noisy, and tiring. Lifting the galley trays was arduous, and working with ink was messy. Working with lead, chemical solvents, and the machinery involved health hazards. Cockburn found that although the men complained of these things, they also felt nostalgic for them, with positive associations derived from the gender connotations of the work. As one of the men she interviewed said,

> I like to do a man's job. And this means physical labor and getting dirty. . . . To me, to get your hands dirty and work is . . . working brings dignity to people. They are doing something useful, they are working with these [he demonstrated his hands]. . . . That's what it is all about. Craftmanship.[65]

Like the workers Zuboff studied, the men Cockburn interviewed seemed to be experiencing "epistemological distress," as well as gender disorientation.

Photocomposition, on the other hand, is "white-collar" work, in that it is clean, quiet, and nonarduous. Although the men appreciated these features of the work environment, they also felt out of place in the new workplace. The officelike environment was perceived as less virile; "less manly somehow," as another photocompositor said.[66]

These feelings of emasculation are also related to actual changes in the labor process. The men had to be taught traditional typing skills: touch typing on a conventional typewriter keyboard instead of sight typing on the wide keyboard. Although the men were learning a new skill to replace the old, they *felt* deskilled. Cockburn argues that this was largely because of the gender definition of typing as women's work, coupled with the fact that the new keyboards were generally too small for male hands. Moreover, touch typing, along with the fact that the process of right-margin justification was computerized, deskilled the work by eliminating conception; as one man put it, "I feel like a sausage machine, taking words in and spewing them out all day long."[67] Another said, "It's taken the soul out of the job."[68]

The men's relationship to the machines themselves also changed. Whereas the workers had a physical and mental relationship with the hot-metal Linotype machines, attuned to its sounds of operation and visible manifestations of needs for maintenance or repair, the VDT terminals were like a mysterious "black box" to them. This technology appeared to need no maintenance, and any repair was done not by operators but by maintenance technicians. The men felt embarrassed not to understand how the machine worked.

The shift from metal to paper has implied not only a change in skill requirements but also a sociopolitical shift in power. Cockburn explains:

> Whose is this labor process? A change in material from metal to paper has led to a shift in power. Lead alloy and the machinery used to process type may have literally belonged to the capitalist, but in effect they belonged to the compositor, who alone knew how to put them to work. Paper and glue—these are the materials of the kindergarten, they are everyone's thing. The journalists and editors who come down to the stone to see their stories being made up used to be unable to read the lead type except with difficulty, back to front as it was. Now they look over the comp's shoulder and can read the bromides as well as he. Composing has lost its mystique and the compositor much of his authority.[69]

Finally, the general work culture has changed. The most obvious aspect of this change is the changing sex composition of the compositor work force. Women are beginning to enter photocomposition, slowly in the United Kingdom, but in the United States, women were 41% of compositors by 1980.[70] The integration of women has upset the traditional male culture of the compositor trade, with its sexism and machismo. The female compositors whom Cockburn interviewed said that they felt like objects of curiosity and disdain. The men missed the male camaraderie of the past and also felt that "if girls can do it, then you are sort of deskilled, really."[71]

Another dimension of the changing work culture is confusion over class. Hot-metal composition was clearly a skilled blue-collar occupation, and yet photocomposition is not. Is it a clerical job, and therefore downgraded in the occupational hierarchy? Or is it another type of white-collar occupation, traditionally viewed as higher in occupational status than blue-collar work? Cockburn calls it "a semi-mental, semi-manual editorial/typographic interface between the origination of written material and the printing press,"[72] but this is a bit cumbersome for purposes of subjective class identification. Understandably, the men felt a bit confused and disoriented. Adding to their confusion was their sense that cold composition represented technological progress and that to oppose it would be futile or backward.

The occupation of photocompositor is probably becoming obsolete, as the trend toward integrating photocomposition with journalistic and editing functions gains momentum.[73] Platemaking also is on the decline, in terms of both numbers of workers and degree of skill required, as pages are photographically reproduced and transformed into a plastic plate through a chemical process, bypassing the manual "paste up" process.[74] Pressroom workers continue to retain such traditional craft skills as controlling the positioning of the plates, the amount of ink, and the proper mixtures of other chemicals, although laser technology, currently under development, would virtually eliminate pressroom work.[75] Finally, mailroom work has also been automated, with automatic stackers, folders, and machine insertion of flyers, approximating a "continuous flow" technology. The number of mailroom workers has been reduced, but the work of the remaining workers has probably been upskilled.[76]

Although initially the organizational impact of computers on the firm was to polarize the overall labor process into expert (professional journalists and editors) and nonexpert (clericals, photocompositors, mailroom workers, etc.) sectors, as Cockburn points out, phase two of the reorganization may involve a more integrated computerized system of information processing that not only links formerly different occupations but often takes the form of their integration in the same person (e.g., a person who writes, edits, and inputs his or her copy). The number of nonexpert jobs is being reduced (although not eliminated) by the nature of the technological innovation. A. H. Raskin, labor analyst for the *New York Times*, says: "Now a handful of executives and confidential secretaries with a modicum of special training can do everything necessary to produce a paper."[77] As Cockburn comments, "small comfort for the majority of compositors discarded from the labor force, but satisfactory enough for those left within the charmed circle."[78]

SUPERAUTOMATION

In some production workplaces, even more complex and highly integrated "state of the art" computer systems link all aspects of the production process: resource planning, process planning, design, manufacturing. Instead of "islands of automation . . . with human bridges,"[79] computer-integrated manufacturing (CIM) links these islands into a highly integrated, superautomated factory. Flexible manufacturing systems (FMSs), "capable of producing a range of discrete products with a minimum of manual intervention,"[80] are being designed that can manufacture up to five hundred different parts with the same superautomated system. These "factories of the future" make possible dramatic reductions in the size of the production work force as well as the automation of certain managerial functions. When combined with robotics, the computerization becomes almost total. The existing superautomated factories, although few in number, demand attention due to their projected growth and prevalence in the future.[81]

The automobile industry, in the face of intense foreign competition and economic crisis and labor conflict at home, has been at the forefront of experimentation with robotics, flexible manufacturing systems, and CIM. Computerization and microelectronics make possible both the decentralized globalization of production and the integration of the different aspects of production into a centralized manufacturing complex, giving corporations new options for dealing with turbulent markets and militant workers.[82] Moreover, FMSs are attractive because of their ability to adapt to product innovation and style changes.

GM's "Saturn model," for instance, designed to produce small cars in the United States competitively with Japan, integrates various computer-based technologies, from design to the various aspects of production, into a highly centralized manufacturing system that begins to approximate complete computerization. In the Saturn system, residual manual assembly workers, "islands" of human labor, work in teams doing modular assembly at stationary work sites, with various portions of the car brought to the assembly line on a computerized cart.[83]

Robots have also been widely used in automobile manufacture; by 1984, the auto companies were buying over half of all robots used in the United States,[84] and GM anticipated acquiring twenty thousand additional robots by the 1990s.[85] Although "smart robots" are most widely used to do unpleasant, repetitive, and easily automated jobs, such as spot welding or spray painting, increasing technological sophistication may widen their range of tasks.[86] Shaiken contends that the early deployment of robots in these undesirable jobs was a function of technological limitation and that "with return of investment as the leading criteria, the second generation of robots will be targeted for some of the most desirable production jobs in the factory—light assembly, machine unloading, and inspection."[87] Moreover, with robots becoming less expensive, they are increasingly cost-competitive with auto workers.[88]

Technological innovation within the auto industry has also paralleled and facilitated the globalization of production.[89] Although the initial "dual strategy" was to export labor-intensive work abroad and to emphasize advanced automation at home, more recently superautomated factories are being exported to newly industrializing countries. Shaiken and Herzenberg intensively studied a highly automated Ford engine factory in Hermosillo, Mexico, and their findings are worth examining in some detail.[90]

This state-of-the-art engine factory incorporates advanced technology from North America and Western Europe with workers from both Mexico and North America. It features an automated parts storage and retrieval system, laser technology for crankshaft drilling, robot welders, and ninety-nine computerized quality checks performed automatically at the end of the line.[91] The plant could have been even more completely automated, but in a few instances, given the low labor costs in Mexico, cost-effectiveness dictated less automated equipment. The overall CIM production system is among the most sophisticated in the world.

Workers in this highly automated factory monitor, inspect, and maintain the technology. There are only six categories of workers: two skilled and four production classifications, as compared with 150 job classifications in a less automated engine factory.[92] Low-level workers are given more training than in comparable North American plants, and promotion from one production category to another is based on managerial decisions about merit (as compared with seniority in more traditional plants). A core of managers and engineers was brought in from various plant locations around the world, and the rest of both skilled and semiskilled production workers were trained locally in a nearby technical institute. Although all categories of workers work collectively to repair breakdowns, managerial authority is strong.

Indeed, Shaiken and Herzenberg contend that a basic reason for the plant's successful operation is its personnel strategy: "The locus of decision making was moved from the young shop floor workers to seasoned managers."[93] To operate and monitor highly automated equipment requires a type of diagnostic skill that is difficult to acquire except through experience; therefore, in this plant experienced managers "routinely make operating and maintenance decisions which are made by operators, skilled workers, and first-line supervisors in a North American plant."[94] Indeed, in this flexible manufacturing system the most salient type of flexibility is managerial flexibility derived from three features of the plant: broad job descriptions, the ability to assign workers to any job, and little demarcation between job categories.

This Mexican superautomated factory achieved efficiency, productivity, and quality comparable to that of its North American counterparts within two and a half years, with wage costs of less than 10% of those of U.S. plants. Based on this performance, Shaiken and Herzenberg conclude that more high-tech factories will be exported in the future. In addition to cheaper labor costs, other advantages include closer proximity to raw materials, incentives given by newly industrializing counties to attract manufacturing, and a work force with "positive" work attitudes.

THE INTERACTION OF SOCIAL AND TECHNICAL FACTORS

Although much of the empirical work on computerization treats advanced technology as an independent variable, it is clear from other studies that technology interacts with social and political factors. Duncan Gallie compared oil refineries with very similar types of continuous-process production technology but in two very different national contexts: France and Britain.[95] His primary research questions concerned the implications of advanced automation for social integration of the work force, the changes (if any) in managerial power, and the implications for unions. He found that "advanced automation proved perfectly compatible with radically dissimilar levels of social integration, and fundamentally different institutions of power and patterns of trade unionism."[96] In contrast to theorists who have attempted to derive general theories of the impact of advanced technology, Gallie argues that technology must be analyzed in social context.

French oil-refinery workers were generally more politicized, with strong desires to participate in workplace decision making; these attitudes were encouraged by their union. In actuality, however, French management was rather authoritarian, and workers had little influence on decision making. This disjunction between desire and reality led to a polarized and acrimonious atmosphere within French plants: "The enterprise was seen by the workers as socially dichotomous, and exploitative."[97]

English workers, on the other hand, were relatively satisfied with existing plant operations. In contrast to the highly centralized French managerial structure, English plants were more decentralized and British workers had a higher degree of participation in decision making; control of everyday work processes was under the jurisdiction of worker teams. British unions, while not as incendiary as their French counterparts, were more successful in bargaining with management and in winning concessions. When British workers expressed workplace grievances, these were more likely to be "technical criticism . . . of the efficiency with which management carried out its duties."[98] British workers were fundamentally committed to managerial objectives of rationalization and efficiency.

Based on these findings, Gallie argues that social and political factors are far more influential in determining the nature of workplaces than are technological ones. He does not deny that technological change may be somewhat influential within a particular national context, but merely that it seems clearly subordinate to social context. Virtually identical technological systems can have very different effects, depending on such factors as the nature of unions, the nature of management, and the way in which organizational control is structured within the plant.

Another timely national comparison that is instructive with regard to the interaction of technical and social factors focuses on differences in material production between the United States and Japan. Recent evidence indicates that while Japan enjoys some degree of technical superiority in some industries, the most important reasons for Japan's competitive advantage are social.[99]

Lester Thurow, for instance, contends that the productivity of U.S. industries has been in decline for twenty years due to "structural weaknesses" such as an emphasis on short-term rather than long-term planning, poor labor/management relations, poor coordination, lack of public investment banks, and underinvestment in research and development (particularly nonmilitary R&D).[100] He argues that because the United States has lacked a viable industrial policy, it has relied on protectionism, which is a futile and misguided strategy. Unless the United States transforms the social organization of work, it has no possibility of becoming competitive in world markets.

A more in-depth analysis of differences in industrial organization between the United States and Japan is provided by David Friedman.[101] He argues that the social organization of factories is highly divergent in the two countries; in the United States the general model is one of standardized production, whereas in Japan the approach is one of flexible production. Standardized production emphasizes Fordist strategies of mass production: producing a given product as cheaply as possible. Specialization of machinery, routinization of work, deskilling of the work force, and resistance to change are the norm. Conversely, flexible production focuses on product and process innovation, leading to a very different type of factory:

> The labor force in flexible firms is very different from a standardized workforce. Several factors make the deskilling required by standardized production undesirable to flexible manufacture. First, the ever-changing product needs are not served by the establishment of workplace routines and the installation of highly specialized machines. Instead, skilled workers on general-purpose machinery are employed to put out rapidly changing goods. Second, any attempt to continuously modify products will inevitably lead to production snafus, but a skilled, autonomous workforce is far more capable of immediately and efficiently resolving production problems than deskilled labor. Finally, a skilled workforce is also a productive one, and a source of ideas that can help promote new designs. Thus, high skill leads both to higher productivity and to increased flexibility.[102]

It appears, then, that technological differences are less salient than differences in the design and implementation of a similar technology.

Friedman uses data drawn from a comparative analysis of the

automotive industry in the two countries to illustrate this basic structural difference. He contends that differences in industrial philosophy
can be seen as early as the post-World War II period, when the big
three U.S. auto makers, wedded to their standardization of product
and process, did not respond to the challenge of developing a small
car, despite widespread interest in one. Japan, on the other hand, did
develop small cars; although its largest producers (Toyota, Nissan, and
Prince) opted to continue making large cars at this time, smaller producers began developing small cars and carving out a new market niche
that has grown steadily since that time.

Even more revealing is Friedman's analysis of recent competition
between U.S. and Japanese auto makers. He argues that since the mid-
1950s, when Japan mounted its export drive, Japan has "continually
attempted to gain market shares through product differentiation, while
U.S. companies stuck gamely to their standardized model lines and
tried to squeeze out foreign challengers by lowering their prices."[103]
During the 1960s, while U.S. firms made minor, cosmetic changes in
its models, Japanese firms were designing completely new models tai-
lored to the American market. Even before OPEC created a fuel-econ-
omy emphasis and a dramatic shift in consumer taste, Japanese (and
other) imports were a significant threat. In 1973, imports claimed 15% of
the U.S. market.[104]

The pricing tactics of U.S. firms were increasingly ineffective, as
Japanese cars were designed to appeal to consumers on other grounds:
performance, styling, technological superiority, fuel economy. By the
end of the 1970s, imports had captured almost 30% of the U.S. market,
aided by the OPEC embargo. Friedman argues that it was Japanese
flexibility that facilitated this strategy of product differentiation and
market fragmentation, a strategy to which U.S. firms have yet to
respond. Indeed, U.S. firms have responded with more of the same: a
continuation of standardization. Homogeneity of product, large volume, and economy of scale continue to be emphasized. The recent
"world car" strategy of GM and Ford exemplifies this philosophy: a
single design, produced on a global scale at the least possible cost.[105]

Friedman argues that the U.S. adherence to standardized strategies, even when they have obviously not been effective, is based on a
fundamental misunderstanding: that efficiency and cost-effectiveness
are diametrically opposed to product differentiation and innovation.
In actuality, however, as the Japanese case aptly demonstrates, flexible production can be cost-effective. One important factor is the greater

skill of the workforce, which leads to enhanced productivity. Indeed, Friedman documents that it is the greater productivity of Japanese workers, rather than their lower wages, that led to the Japanese competitive edge.[106] Moreover, in addition to being more productive, Japanese workers are also more creative and more able to solve unanticipated problems effectively, factors that also contribute to quality production. Japanese management practices, such as quality circles, are designed to develop these qualities in workers. Conversely, even though U.S. firms have sought to emulate Japanese management, these attempts have largely failed, due to the underlying emphasis on standardized production, which is antithetical to Japanese management.

Hirschhorn has also analyzed the interaction between technology and the social organization of production at firms such as Olivetti, Fiat, General Foods, and other companies.[107] He found that innovations along the lines of sociotechnical proposals (e.g., self-governing worker teams, an emphasis on worker learning, job rotation, coordination rather than managerial control) resulted in better quality products, expanded worker control, and superior product innovation and market position.[108] Conversely, in factories where centralized managerial control is combined with worker deskilling and alienation, deskilled workers have had difficulty effectively monitoring and diagnosing the sophisticated equipment and its breakdowns. He analyzes the case of Three-Mile Island as an example of how deskilled workers were unable to successfully interpret the signals of complex technological failure so as to correctly diagnose the situation. Given the complexity of the technology, the possibilities for failure are similarly complex. Overspecialization is another impediment to comprehensive understanding and correct diagnosis of problems.

It seems clear that social and political contexts have dramatic effects on how a given technology is designed and implemented.[109] In the United States, it seems that an underlying emphasis on control, efficiency, and mass production can undermine the positive potential of advanced technology and innovative workplace organization. Sociotechnical analysts have stressed the importance of more comprehensive worker knowledge and ongoing worker learning if the potential of advanced technology is to be realized. The empirical studies reveal that although some firms emphasize training and learning, this is often of a rather circumscribed sort.[110] Technological innovation may open up new opportunities for workplace restructuring, but it does not appear to be sufficient to insure either workplace democratization or enhanced worker autonomy and learning. In fact, it appears that social and political factors often override technological ones, producing contradictory outcomes.

CONCLUSION

Clearly, blue-collar production workplaces have been undergoing dramatic changes during the past few decades. Substantial variation is apparent, but underlying patterns and parameters of organizational change can nonetheless be discerned.

In terms of the debate between neo-Bravermanians, who allege that advanced technology tends to be deskilling and Taylorist in its effects, versus the sociotechnical analysts (and others) who contend that computerized systems tend to promote workplace democratization, upskilling, and worker autonomy, the empirical evidence reveals that each perspective contains both truth and limitations.

For many reasons, skill is a complex issue. In a context of rapidly changing job classifications, whether a job is deskilled or upskilled depends in large measure upon the skill levels of one's *previous* job. As we have seen in the newspaper industry, photocompositors appear to lose certain craft skills in the shift to photocomposition, whereas mailroom workers experience upskilling as a result of continuous-process automation of the mailroom. Numerical control erodes traditional machinist craft skills but also opens up opportunities for reskilling through end-user programming.

Working with advanced computer systems requires fundamentally different skills, skills that sociotechnical analysts contend demand more comprehensive worker knowledge, team work, and autonomy. However, in many workplaces, workers are not given such advantages, even at the price of suboptimal performance and thwarted technological potential. A central reason appears to be managerial resistance to change and reluctance to share power.[111]

Moreover, skill levels are not necessarily correlated with degree of worker autonomy, although both neo-Bravermanians and sociotechnical analysts tend to make this assumption. Workers can be relatively well trained and skilled and yet subject to control by managers or technical experts.[112] During a time of rapidly changing skill requirements, which skills one has becomes important. Ability to understand the technological system, to interpret the information it provides and diagnose its failures—these become the skills that are in demand and that are rewarded. And yet workers are often not given the sort of training that would allow them to acquire such skills.

Skill levels also may differ according to the level of analysis: task or job. A given *task* may be deskilled, but depending on the task specifications of the job, the *job* may be upskilled. As we have seen, job rotation and teamwork provide alternatives to fragmentation and worker isolation, mitigating any deskilling tendency through task variation. However, whether a job that involves several deskilled tasks is appreciably more skilled, or merely speeded up, is open to debate.

Skill is also transformed due to the fact that advanced technology tends to demand more white-collar skills. Blue-collar production workplaces are becoming more similar to white-collar workplaces: less arduous, less messy, centered around sophisticated technological systems that demand more abstract and conceptual skills.[113] As Cockburn has shown, this shift has had dramatic implications for class identification, gender relations, and perceptions of skill levels.[114] Male production workers now working with computerized technology may feel less masculine and deskilled, although in fact new skills have been acquired. Subjective evaluations of skill may be influenced by gender assumptions and may differ from objective evaluations the changes in skill requirements.

The erosion of craft skills implies that the relationship between conception and execution has been further altered. Not only has execution been deemphasized in favor of monitoring, but in some workplaces the gap between conception and execution has widened, as craft and supervisory positions are deemphasized.[115] The conceptual sector becomes an expert sector, composed of both high-level managers and technological experts, and the nonexpert sector is reduced in size. Expert and nonexpert sectors are therefore clearly defined; Shaiken speaks of a "two-tiered workplace: a small number of creative jobs at the top and most other jobs with fewer skills and subject to new forms of electronic monitoring and control."[116]

However, there is also evidence that in other workplaces the conception/execution gap may be in the process of being reduced, as expert workers (e.g., journalists) perform tasks once allocated to nonexpert workers, and as nonexpert workers (e.g., workers in superautomated factories) take on certain conceptual tasks and work more cooperatively with technical experts. These changes raise further questions about nonexpert-sector worker displacement and the need for retraining.

There is, then, considerable complexity, ambiguity, and variation surrounding the transformation of skill levels. It is clear, however, that certain craft and manual skills are being rendered obsolete, and new skills (monitoring, diagnosis, maintenance) emphasized, due to both technological and social factors. What Randy Hodson terms "skill disruption"[117] is certainly occurring, and given the fact that most transformations of requisite skill are foreseen or planned, "skill restructuring"[118] is probably a more precise term.

In general, technical skill and expertise are becoming more important, at the same time as traditional craft skills are being eroded and traditional managerial skill is being transformed. Understanding the technology itself, which is less visible and less readily comprehensible in its operation than mechanical technology, as well as the ability to interpret the large amounts of technical information that computerized technology makes available, are increasingly important skills. Information is power—at least to those who can effectively utilize it.

> The information-gathering potential of the technology is impressive. . . . In the latest systems, the collection and transmission of data occurs as the event is taking place, literally at the speed of light. In the past, reports of shop floor activities were transmitted by verbal command and written memo. . . . Information more or less drifted up to the desired level. Now production problems are relayed instantaneously to the designated point . . . and routine information is compiled on a much more frequent basis.[119]

Which information is made available, to whom, and how it is used— these are increasingly important variables, and the focus of political struggle between management, technical experts, and workers.

The effects of technological change are also ambiguous due to the importance of social and political context. As we have seen, similar technology can have very different effects in different contexts. Microcomputer technology, for instance, can promote decentralization and end-user programming or enhanced managerial surveillance and monitoring of workers. Differences in national context, degree of unionization, and the balance of power between management and workers are all influential in shaping technological implementation.

Gender is another social variable that shapes technological change. As certain predominantly male crafts (e.g., composition, garment cutting) are transformed, and as more arduous blue-collar work becomes more white-collar, feminization is occurring. It appears that in many workplaces the breach between expert workers (programmers, computer engineers, maintenance technicians) and nonexpert workers is paralleling and reinforcing sex segregation. Cockburn concludes from her study of eleven computerized workplaces:

> When the dust settles after the technological revolution, the same old male/female pattern can be seen to have re-established itself. The general law seems to be: women may press the buttons, but they may not meddle with the works. . . . Always the person who knows best, who has the last say about the technology is a man.[120]

Technological change and workplace restructuring are, then, highly politicized phenomena. The wide range of conceptualizations and empirical evidence concerning these changes point to the range of social and political choices available.

Chapter 4

THE TRANSFORMATION
OF BUREAUCRACY

Bureaucracy arose in conjunction with industrialization, a complementary control structure instituted to rationalize the administration of the expanded production apparatus.[1] Given their shared history, it is not surprising that bureaucracies have changed in recent years, in ways that parallel changes in production workplaces. As we have seen, recent decades have witnessed a certain convergence of blue-collar and white-collar occupations. What are the similarities and differences between blue-collar and white-collar workplaces? How is technocratic restructuring transforming bureaucracies?

Advanced technology, the increased salience of technical and professional experts working within bureaucracies, an increasingly competitive and unpredictable world economy and enhanced demand for innovation, and the development of permanent planning staffs have typically been cited as reasons for changes in bureaucracies. Various new types of bureaucracies have been proposed. Ilchman, for instance, delineates a "rational-productivity bureaucracy," Gouldner a "technical bureaucracy" or "scientized bureaucracy," Goss an "advisory bureaucracy," Smigel a "professional bureaucracy," Mintzberg an "adhoc-racy," and Larson a "technobureaucracy."[2]

Most analysts have seen these changes as modifications in a basic structure that remains essentially bureaucratic; for instance, as "hybrid forms of organization that deviate from the bureaucratic model in order to accommodate their professionals."[3] However, given the substantial changes that have occurred, not only as a result of the incorporation of professionals into bureaucracies but also because of the impact of sophisticated technological systems and related socioeconomic shifts, a new paradigm for understanding the transformation of bureaucracies is needed. This chapter analyzes recent theoretical and empirical work on changes in bureaucracies in order to point toward such a reconcep-tualization.

CONCEPTUALIZATIONS OF CHANGING BUREAUCRACY

As professionals and technical experts have increased their num-bers and influence in most bureaucratic organizations, professional/bureaucratic conflict has been the focus of renewed atten-tion. Earlier analyses of the structural effects of such professional/bureaucratic conflicts focused on organizations where professionals were well represented, such as hospitals, and the ways in which bureaucratic structure was modified to accommodate profes-sionals.

Mary Goss, for instance, analyzed the nature of the authority to which physicians within a large medical bureaucracy were subject.[4] She found that in certain areas (administrative concerns) physicians were subject to bureaucratic control mechanisms, such as scheduling rules, although these rules were often implemented flexibly in order to permit physicians to follow professional norms (e.g., to deviate from the sched-ule when a patient emergency arose). Other areas (professional activi-ties) were exempt from bureaucratic control and were supervised in an "advisory" manner by other professionals, in accordance with norms of professionalism. Similarly, Gouldner's scientized bureaucracy is char-acterized by two separate systems of control: the bureaucratic and that of the technical intelligentsia, with the latter being even more autonomous and exempt from bureaucratic rules.[5]

Magali Larson goes further in analyzing the ways in which "tech-nobureaucratic" modifications of *both* professionalism and bureaucracy are necessary to make them more or less compatible.[6] Externally sanctioned expertise (credentials and recognition by professional organizations) gives professionals some countervailing power within the bureaucratic hierarchy, power that enables professionals to exercise expanded autonomy and discretion in the pursuit of professional goals. Professionals are segregated from other members of the bureaucracy and enjoy special privileges. However, bureaucratized professionals also must modify their traditional client orientation so as to incorporate bureaucratic goals of efficiency and profit maximization (in corporate settings), and they also tend to focus upon more specialized skills rather than the broader mystique of traditional professionalism (see also chapter 6).

In one of the fullest elaborations and analyses of changing bureaucratic structure, Henry Mintzberg distinguishes between *machine bureaucracies, professional bureaucracies,* and *adhocracies.*[7] Machine bureaucracies are similar to the Weberian ideal type: standardized responsibilities, qualifications, channels of authority and communication, and task specifications. They emphasize centralized control, a strict division of labor, and routine cases. Their major limitation, according to Mintzberg, is that they are "fundamentally nonadaptive structures,"[8] ill suited to complex and changing environments.

Professional bureaucracies, such as universities, hospitals, and school systems, are more decentralized but also standardized, emphasizing standardization of skill, training, and professional indoctrination as alternatives to centralized control. External, professional authority is invoked in order to complement professional autonomy; administrative staff function largely to support the professional work that is the main emphasis of the organization. Mintzberg does not consider professional/bureaucratic conflict, assuming that professionals are given sufficient autonomy to minimize any conflict; in fact, he assumes a high degree of decentralization and democratization, combined with such limitations as problematic coordination, innovation, and adaptation to new circumstances.

Adhocracies, in contrast, are organic, adaptive, and innovative. They do not rely on standardization or formalization, and they have a more fluid division of labor and a constantly changing internal structure. Teams of experts that change their internal composition in response to specific projects and changing organizational needs are the central organizational feature of adhocracies. Experts are given considerable power and autonomy; managers are typically experts too, integrated into project teams and serving a coordinating function rather than traditional managerial control functions. Thus, "administrative and operating work tend to blend into a single effort . . . an organic mass of line managers and staff experts . . . working together in ever-shifting relationships and ad hoc projects."[9] Adhocracies are particularly well suited to complex and dynamic environments, and Mintzberg clearly feels that this organizational form will become increasingly prevalent in the future:

> This is the structure for a population growing ever better educated and more specialized, yet under constant exhortation to adopt the "systems" approach—to view the world as an integrated whole instead of a collection of loosely coupled parts. It is the structure for environments becoming more complex and demanding of innovation, and for technical systems becoming more sophisticated and highly automated. It is the only structure now available to to those who believe organizations must become at the same time more democratic yet less bureaucratic.[10]

A similar but less well-known analysis is provided by Warren Ilchman.[11] He presents a reformulation of bureaucracy that is more grounded in the changing sociopolitical context and more focused on the increased salience of technical rationality within bureaucracies.[12] He postulates a new type of "rational-productivity bureaucracy" as a more rationalized, contemporary modification of Weber's ideal type of bureaucracy. Ilchman finds that many of the factors that Weber specified as responsible for the emergence of bureaucracy, such as the increased size and dominance of the state, an increasingly complex economy, and corresponding needs for planning and technical rationality, have intensified.[13] All of these factors, along with the increased emphasis on education and changing technologies, imply a new type of rationality and new needs for productivity, as opposed to the legal rationality and efficiency orientation of Weber's bureaucracy.

Specific ways in which a rational-productivity bureaucracy differs from a legal-rational bureaucracy include:

1. fewer hierarchical levels, and less importance given to hierarchy
2. decreased relevance of formal monocratic authority
3. decreased relevance of formal communication channels
4. "ambivalent" centralization, with productivity as the governing factor
5. increased importance of preentry education and training
6. increased emphasis on normative compliance
7. increased emphasis on interorganizational mobility

8. knowledge rather than legal authority as the main source of legitimacy
9. expertise rather than office as main source of power
10. decreased functional specialization
11. loyalty to profession rather than to organization
12. a time orientation that is more future oriented
13. increased capacity to change

Although Ilchman is primarily concerned with using this ideal type as a framework for analyzing how rational-productivity bureaucrats have fared in developing countries, this typology of contemporary changes in bureaucracies has broader implications.

In order to further explore contemporary changes in bureaucratic structure, we turn now to an examination of some of the recent empirical literature.

HIGH-TECH CORPORATIONS

Given their emphasis on advanced technology, technical experts, and innovation, and their competition in dynamic, unpredictable environments, high-tech corporations might be expected to be at the forefront of organizational change. Several recent studies have examined the structure of high-tech firms and how it differs from more traditional bureaucratic organization.

Kanter, for instance, studied upper-level employees and organizational structure in five high-tech corporations.[14] She also compared six high-tech, socially innovative corporations with four more traditional, bureaucratically organized companies.[15]

Kanter found that the high-tech firms she studied were relatively young and marked by rapid growth and a need for continuous innovation in order to remain competitive in foreign markets. The recruitment, motivation, and retention of technical talent were therefore crucially important for the attainment of organizational goals, leading to changes in bureaucratic structure. In fact, Kanter contends that such needs for innovation and technical expertise not only are inherent in high-tech firms, but are increasingly characteristic of corporations in general. However, in her sample, the high-tech firms differed dramatically, in terms of organizational innovation, from the more conventional industrial and service corporations.

Relatively young and rapidly growing firms that need to recruit technical personnel necessarily emphasize external labor markets and credentialing. Kanter found that the younger high-tech firms in her sample were particularly likely to fill non-entry-level positions from the outside, and that all of them relied on external recruitment of technical personnel a great deal. This reliance is partly a consequence of the inability of on-the-job training to adequately train technical experts, and partly a consequence of rapid growth: "Unpredictable or too rapid growth does not permit advance planning or waiting for the slow internal development process to occur."[16]

In order to motivate and retain technical workers so as to insure product innovation, high-tech firms rely on a variety of organizational innovations. The overall structure is more decentralized, with a tendency toward larger numbers of relatively autonomous divisions and matrix organization, whereby managers and technical workers report to two or more bosses (one for each functional specialty or market area, for instance). Interdisciplinary task forces and special project teams are constantly created and re-created to work on product development or technical problem solving. In general, the authority structure is much looser, with decentralized decision making, expanded discretion (for technical workers), and an emphasis on horizontal communication and responsibility as alternatives to vertical control.

In contrast to the more formalized organizational structure of conventional bureaucracies, with its clearly defined chains of responsibility, internal job ladders, and vertical communication channels, these high-tech firms relied less on the formal organizational structure and more on "dotted-line relationships, which are often not shown on the charts."[17] Technical workers were more likely to be organized in ad hoc teams and task forces and to make nontraditional moves (both horizontal and vertical) based on particular needs of the firm or exhibited technical expertise. The task-force structure, which was described as resembling a guild, was a way of overcoming the limitations of overspecialization and providing both learning and career opportunities for technical workers.[18] The overall ethos is "integrative" (as compared with the "segmentalist" nature of traditional bureaucracies):

> The willingness to move beyond received wisdom, to combine ideas from unconnected sources, to embrace change as an opportunity to test limits. To see problems integratively is to see them as wholes, related to larger wholes, and thus challenging established practices—rather than walling off a piece of experience and preventing it from being touched or affected by any new experiences.[19]

The team structure was not without its problems, however. Hierarchy was in some cases not readily overcome, and teams composed of members with different statuses sometimes "slipped into deference patterns which give those with higher status more air time, give their opinions more weight, and generally provide them with a privileged position in the group. . . . teams . . . may end up duplicating the organizational hierarchy in miniature."[20] Another source of inequality is differential amounts of knowledge and information. Kanter found that participation in groups with heterogeneous amounts of knowledge and skill sometimes resulted in the less knowledgeable members feeling "dumb," feeling powerless and humiliated, and feeling dissatisfied with decisions that they supposedly helped to make. For instance, at one firm, secretaries were one group that refused to continue to participate in section meetings for these reasons. Finally, team members differed in verbal ability, personal attractiveness, and interest in the task, leading to differential ability to participate. For all of these reasons, internal group politics were influential, at times becoming a "tyranny of peers."[21]

Management is also transformed in high-tech firms. One way of motivating and retaining technically skilled workers is by giving them managerial status but without traditional managerial responsibilities of supervising other people. The title *manager* comes to imply only high status, with "the manager/non-manager distinction more often used to signify the difference between high-level salaried and hourly employees than it is to clearly differentiate managers and professionals."[22] In effect, it appears that managers and professional/technical workers are organizationally grouped together in an expert sector that is clearly distinguished from the nonexpert sector. Expert sector workers are then given status recognition and privileges that are analogous to those given in academic settings: sabbaticals, fellowships, and so on.

Another way in which management is changed is that traditional forms of authority are no longer operative. In matrix organizations, where employees and managers alike report to various people, the unitary chain of command is abolished. Kanter argues that *influence* substitutes for authority as a means of gaining compliance: "Traditional authority virtually disappears; managers must instead persuade, influence, or convince,"[23] creating adjustment problems for managers accustomed to exercising authority based on rank position. Other studies have confirmed the increased salience of technical expertise among managers of high-tech firms and the decreasing influence of rank authority.[24]

Kanter found that, despite relatively advantageous working conditions, technical workers typically showed little company loyalty and often used "company hopping" strategies to increase their salaries and organizational position. This was partly because their skills were more generalizable and less company-specific and partly because of the flatter opportunity structure of high-technology firms. Geographic mobility was also emphasized, both interfirm and intrafirm: "High-tech operations are spread across the country—and the world—so as to not exhaust the technical talent in any one area."[25]

Working conditions for nonexpert workers in high-tech firms were more rigid and conventional, with autonomy and innovation emphasized less than reliability and "turn-it-out-on-schedule."[26] In some settings, Kanter found mechanistic and "organic" types of organization coexisting.[27] In other firms, attempts were made to energize and involve the grassroots, for instance, through ongoing nontechnical education, action projects aimed at solving company problems, and worker teams.[28] Sometimes organizational innovation at the nonexpert level was substantial and democratizing, as in the case of action project teams that designed and instituted a new team structure for themselves; in other cases it was superficial, as in the example of a manager, surprised to learn of worker dissatisfaction, who suggested making a "Trust the System" T-shirt.[29] In some companies, the wide breach between expert and nonexpert sectors contributed to pronounced difficulties in the sectors' comprehending each other's work issues.

In another study, of twenty-two high-tech companies, that included interviews with managers, engineers, and workers, Randy Hodson has analyzed the changing organizational structure of high-tech firms and the implications of these changes for workers at all levels of the organization. He found that "there is substantial evidence that high-tech firms are experiencing quite serious, even if hidden, organizational crises."[30]

One of the most universal organizational changes that Hodson noted was a polarized segmentation of the organization, with "very constrained intrafirm mobility opportunities for workers,"[31] an emphasis on lateral moves rather than promotions, and a trend toward increasing income inequality and segregation by race and sex. Production and clerical work forces had higher concentrations of women, minorities, and young workers; engineers were virtually all white and 80% to 85% male. Both women and racial minorities complained about discrimination and blocked opportunities, and turnover was high at all levels of the organization.[32]

Working conditions for engineers involved high levels of autonomy and challenge, craftlike working conditions, "a high degree of both
formal and informal participation in production decisions," close-knit
and well-functioning project teams, and "embarrassingly high
salaries."[33] Although the engineers sometimes complained of managerial incompetence to adequately supervise their work, given managerial
distance from the production process and lack of technical knowledge,
and of managers' preoccupation with bureaucratic rules and marketing,
the engineers nonetheless were allowed relative autonomy to conduct
their work in accordance with craftlike norms.

Nonexpert workers, whether production workers or clerical workers, were typically not genuinely involved in work teams or decision-
making processes: "Workers in large companies are asked to 'participate'
in superficial and facile ways that are a thin substitute for more genuine
forms of participation, such as those existing for engineers and, to a certain
extent, even for workers in smaller companies."[34] Workers often felt that
managerial control was abusive and arbitrary, as well as incompetent.
They also complained of underutilization and inadequate opportunity.
However, workers liked the more informal work atmosphere and the
very fact that they were working with advanced technology.

Although Hodson found that internal labor ladders were largely
nonexistent in the more polarized organizational structure, he also
found that on-the-job training and firm-specific skills were still quite
important, particularly for nonexpert-sector workers. Managers said
that the most important criterion in the hiring of low-level workers was
"attitude." Among engineers, a combination of prior training and on-
the-job training were needed. This ongoing need for on-the-job training,
combined with the fact that turnover rates were high among workers
and engineers alike, led to a crisis of organizational structure and motivation, particularly among nonexpert workers. In order to deal with
labor problems, temporary workers were often used as clerical employees, and subcontracting and offshore production were increasingly used
for production work.

Hodson, then, found organizational changes that parallel some
of the changes in computerized factories (see chapter 3), particularly
polarization and crises of nonexpert-worker motivation. The erosion
of internal labor markets, in conjunction with the ongoing need to train
and motivate workers, was a particular source of contradiction. It
appears that social changes have not kept pace with technological ones
in the firms that Hodson studied.

Glenna Colclough and Charles Tolbert used quantitative analyses of census data to yield a more macrosocial understanding of high-tech occupations in the United States.[35] They found that high-tech industries, more than other occupational sectors, were characterized by a two-tiered occupational structure.

> Economic inequality in high-tech industries derives largely from a two-tiered occupational structure that differentiates professional and technical workers from operative and assembly workers. Off-shore sourcing and capital mobility may reduce numbers and proportions of production workers in particular locations, reducing economic inequality levels in those areas. But rather than eradicating disparities, this strategy exports the inequality between the top and bottom occupational levels to other locations . . . with less permeable opportunity structures, economic inequality in high-tech industries becomes tied to gender, race, and ethnic segmentation more so than in other industries. Similarly, inequality is exacerbated by the increasing use of ancillary production workers in high-tech industries. . . . Moreover, deskilling and certain applications of automation restrict mobility opportunities for the bottom tier of production workers.[36]

Colclough and Tolbert conclude that "a sector that is on the cutting edge of technical innovation appears to be in the Dark Ages in its managerial strategies and organization of work."[37]

SERVICE CORPORATIONS

Kanter predicted that many of the organizational changes that she documented in the high-tech firms she studied would tend to become more generally characteristic of corporations.[38] Recent studies of service corporations indicate that some of the organizational innovations that we have seen in high-tech firms are also apparent in more mature service corporations.

Thierry Noyelle presents detailed case studies of Macy's, AT&T, and a large insurance corporation, focusing on changes in organizational structure.[39] One of his main findings was a significant internal polarization of all three firms. In contrast to earlier hiring structures which had included multiple entry ports, Noyelle found that all three companies had shifted to a "two-tier hiring structure," with a four-year college degree serving as a credential barrier between the two levels. Moreover, he found bimodal salary structures and virtual elimination of any job ladders between the two organizational levels. This was in direct contrast to previous reliance on internal labor markets; before the 1970s at AT&T, for instance, "with seniority, experience, and on-the-job training, rank-and-file employees could expect to reach middle-management, if not upper-management positions," although race and sex barriers were also operative.[40]

Another aspect of this polarization is the elimination of middle-level positions. At AT&T, for instance, computerization has implied reduced need for certain maintenance technicians (with the centralization and automation of much maintenance work) and installers. At the insurance company, computer automation has implied the gradual elimination of middle-level clerical positions, along with semiprofessional positions such as raters and underwriters. Low-level jobs such as stock clerks, telephone operators, and clerical workers have also been reduced in number. At the same time, high-level professional and technical workers, such as computer professionals, have increased in number and importance and tend to be recruited from the external labor market using credentials as a central criterion.

At the large insurance company, the polarization was also related to the regionalization of the firm.

> The increasing emphasis on formal credentials at various grade levels, the disappearance of traditional bridge skills such as those of rating and routine underwriting, as well as the geographical separation of the regional processing centers from the home office, have all acted to bring about a two-tier structure, as suggested by executives in the company itself. Regionalization has reinforced the process because the tendency has been to move clerical processing jobs to the regional center and maintain many of the junior professional jobs at the head office. . . . The capacity for the company to run regional centers with a low ratio of supervisors to production personnel has in part been helped by the use of quality-circle management methods.[41]

This regionalization implies a certain degree of decentralization, in direct contrast to the heavily centralized bureaucratic structure that was characteristic of this company (and most) until the 1960s, but also an overall polarization of the system.

At this insurance company, the experimentation with quality-circle management methods has been combined with experimentation with job rotation and worker teams, organized around specific accounts rather than a more specialized division of labor. The company had tried a more conventional assembly-line specialization of function but found that, with advanced data processing technology, to separate data entry from data processing resulted in higher error rates and worker boredom. The reorganization around teams responsible for a small number of specific accounts also facilitates more personalized and informed servicing of these accounts.

Another reason for the experimentation with more participatory forms of work has to do with worker motivation. With the demise of internal job ladders and the corresponding erosion of opportunities for upward mobility, service corporations are experiencing the need for alternative means of motivating clerical and other nonexpert-sector workers. As one manager at Macy's put it,

> We used to promise upward mobility, but we do not any longer. We have to be able to give something else. We have tried quality control circles in some of our stores, but we haven't really tried very hard. . . . We need much greater involvement of sales clerks. There is a malaise which needs to be alleviated; we need to boost productivity, but people are bored with their jobs and with the prospect of staying thirty years in the same job.[42]

It appears that whereas all three of the companies Noyelle studied exhibited the polarization into a two-tier structure and the corresponding erosion of internal job ladders between the two main levels, companies have dealt with the ensuing personnel implications differently. The insurance company experimented with quality circles, job rotation, and team structure in order to enhance work quality and worker motivation; Macy's managers had some recognition of the need to motivate workers differently so as to reduce their high turnover rates, but their experimentation with different forms of work organization had been limited; AT&T relied on a more centralized system (facilitated by computerization) to control hiring, training, and supervision of a reduced clerical/operator work force.

Among these changing service corporations, the degree and nature of centralization is variable. AT&T continues to use computerized telecommunications technology in a highly centralized framework; Macy's has experimented with an "orbit" system, with some functional decentralization of authority among branch store supervisors; the insurance company not only has reorganized around functional decentralization and regional product specialization, but has experimented with some decentralization of authority, more participatory forms of work, and quality-circle management methods.

Noyelle contends that whereas the early stage of computerization resulted in an intensification of task fragmentation and deskilling, more recently the possibility of a second stage (in part facilitated by micro-computers) has emerged, as evidenced in the insurance company.

> The industry has now entered a stage in the division of labor that lies 'beyond Taylorism'. . . . Historically, the work process in the insurance industry had been organized along an assembly-line principle. . . . If any-thing, the tendency during the 1950's and 1960's was to further parcel and fragment work. The introduction of early computerized systems did not alter dramatically such a division of labor; it simply added one more fragmented task to the sequence—the keypunching of the data needed for computer processing . . . when carriers introduced large-scale distributed systems in the 1970's, they first tried to replicate the old manual proce-dures with the new systems. In recent years, however, insurers have dis-covered the new technology permits them not only to automate many traditional tasks but also to aggregate many tasks into new combina-tions—in short, to move beyond the old assembly-line concept.[43]

Barbara Baran, in an extensive study of the insurance industry, also concludes that recent organizational changes point toward "an electronic transcendence of the assembly line."[44] In contrast to the first wave of computer automation during the 1970s, which typically reflected and perpetuated the existing hierarchical division of labor, new integrated systems are facilitating a more fundamental restructur-ing of the organization of work. Underwriting and claims support sys-tems, for instance, increasingly integrate the entire process around smaller numbers of workers working with the computerized system. Increasingly sophisticated computer software is substituted for skilled underwriters for rating and risk assessment. Accordingly, skilled cleri-cal workers, with the aid of appropriate computer software, assume virtually exclusive responsibility for underwriting: screening initial client information, inputting this data to the computer, evaluating the computer-generated rating of the policy, and issuing the policy.[45] Task fragmentation and the division of labor are therefore transformed, lead-ing to more personalized service as well as to skill upgrading for cleri-cal workers (but deskilling from the point of view of underwriters).

Baran argues that this form of reorganization is dependent not only on the stage of technological development, but also upon managerial policies and choices and upon the nature of the service that is being offered. Insurance underwriting is characterized by high volume (making automation cost-effective) and is sufficiently standardized to make computer automation feasible for most cases. However, the overall process also involves semiskilled functions (largely, interpersonal interaction) that resist more complete automation. Alternatively, Baran says,

> where products are standardized and volume is high but semiskilled functions *can* be assumed almost entirely by the computer system, the work force is highly bifurcated between a large number of routine data-entry clerks and a very small number of skilled professionals. At the clerical level, there is little real decision making and no interaction with agents or clients.[46]

As computer systems become even more sophisticated, this pattern of organizational restructuring could replace the integration around skilled clericals.

Baran concludes that aggregate skill levels have risen in the insurance industry in recent years, as the number of professional and technical workers has risen and clerical employment has fallen, particularly routine clerical employment. Moreover, the new integrated systems can serve to minimize the gap between mental and manual labor, as routine functions such as data entry are "slipped unobtrusively into the activity of higher-level workers."[47]

Rising skill levels do not necessarily imply improved working conditions, however. Baran found that the skilled clerical workers were poorly paid, and in some instances their work was monitored and paced by the system, leading to heightened stress.[48] Feldberg and Glenn conclude from their study of similar developments in insurance firms, "The clerk has more task variety than previously, but operates in a more rationalized system in which each part of the job must be done exactly according to specific patterns and within a specified time frame."[49]

Moreover, there is clear evidence that although skill levels may be rising, mobility prospects are being eroded. One manager in Baran's study, for instance, spoke of a "quantum leap" between the computer-linked clerical positions and the next step up the occupational ladder.[50] As middle-management, supervisory, and underwriting positions are eliminated, a new polarization between upper management/technical workers and skilled clerical workers appears to be emerging.

Zuboff also found that the computerization of service corporations can facilitate either a polarization of clerical work from managerial work or an integration of the two, depending on whether technological innovation is implemented with an automating strategy or with an informating strategy in mind.[51] Her findings from the service corporations she studied indicate that automating strategies and polarization are the norm, largely due to managerial reluctance to lose prerogatives.

Zuboff found that information and data have become important currencies, which are carefully guarded. At one pharmaceuticals company, for instance, a computer conferencing system was instituted for expert-sector workers, with access being limited by passwords, account numbers, and designations of "closed status" for important conferences. For the workers who participated, the conferencing system represented an "opportunity to extend and elaborate the oral culture in which they conducted their professional work,"[52] as well as opportunities to demonstrate knowledge of one's subject area so as to gain new power and influence. Enhanced collegiality was the result.

Conversely, Zuboff found that nonexpert-sector workers felt more isolated when working with computerized systems than previously. Clerical workers spoke of a less social environment, as increasingly they interacted with their own computer terminals rather than with co-workers. One benefits analyst, for instance, spoke of her feelings of isolation in this way:

> I think we're all more separate now. Before, there was more that I could help someone out with. There's not really that much I can do now. You just don't seem to get to know people . . . because you don't interact. The office has become much more impersonal. . . . The girl who pays the Consolidated Underwriters' claims sits right in front of me. There was a question on my claim form. She didn't turn around and ask me. She sent me a letter. She didn't realize it was me. I said, "Cindy, do you know that you sent me a letter?" She said, "Did I really?"[53]

Nonexpert workers also spoke of decreased communication between themselves and managerial workers, saying that "management has become much more distant" and "there are things that they used to have to depend on us to find out. Now they can go right in the machine and find out anything they want. They don't have to communicate as often."[54] Clerical workers also spoke of feeling tied to their terminals and complained of bodily suffering (backaches, eyestrain, etc.).

Zuboff concludes that computerization has typically been implemented so as to enhance centralized power and managerial control. This has been accomplished by making nonexpert workers more visible and vulnerable to supervision, and by protecting expert sector workers from scrutiny—thus creating a technologically advanced version of the Panopticon.

> The Panopticon's genius was that illumination and visibility provided the possibility of total control. . . . The Panopticon represents a form of power that displays itself automatically and continuously. Panoptic power lies in its material structure and presence; the Panopticon produces the twin possibilities of observation and control.[55]

This centralized control is not technologically determined, however; in fact, Zuboff contends that computerization is equally compatible with decentralization.

> The electronic text exists independently of space and time. When text is confined to concrete objects, such as books or pieces of paper, it generates pressures for centralization (you must go to the text if you want to read it) or for possession (you can own the book or maintain your own files). The electronic text is the result of an even more radical centralization: a wide range of information can be gathered and codified in a single computer system. However, this radical centralization enables an equally radical decentralization: in principle the text can be constituted at any time from any place. The contents of the electronic text can infuse an entire organization, instead of being bundled in discrete objects, like books or pieces of paper.[56]

In practice, however, this potential has not been widely realized, and information and data have typically been centralized and carefully guarded.

It seems clear that the majority of service corporations have not moved beyond the centralized, assembly-line organization of clerical work, and that more progressive models of office computerization are far from inevitable. Machung studied the impact of word processing technology in various corporations and concluded that the reintegration-of-tasks model (using decentralized teams) was the exception rather than the rule.[57] In some companies, such as Citibank, task diversity among clerical workers was the goal, but Machung found that in most companies advanced technology has been used to streamline and extend the division of labor.

> Word-processing technologies . . . enable a relatively uncomplicated task, that of typing a letter or manuscript, to be broken down into yet smaller components. In word-processing centers, for example, the different components of typing a letter are divided up and assigned to different people: a supervisor to schedule and allocate the work; a word-processing technician to key the material into the central memory; a printing operator to monitor the output process; a proofreader to catch mistakes; and a clerk typist to pick up and deliver the work. In this sense, word processing closely resembles classical de-skilling: a whole craft is broken down into ever smaller components, mental work is separated from manual work, and bit pieces of the remaining jobs are parceled out to "detail workers." Because of this, word processing centers in large organizations clearly resemble assembly lines in the factory.[58]

The National Research Council concludes from their survey of computerized automation in diverse white-collar settings that "because innovations can be implemented in broadly different ways, the major determinant of the effects of innovation appears to be management's preexisting employee policies."[59]

It appears that firms that adhere to a more centralized model of computer automation also tend to rely on a more specialized division of labor and more stringent technical control. Mary Murphree, in her study of the changing roles of legal secretaries, analyzes two distinct models of word processing: centralized word processing centers and decentralized satellite stations.[60] She found that word processing centers were more "factorylike," in that work was paced and monitored by the centralized computer system, with supervisors using the technology to electronically inspect any terminal's activity at any time. Satellite stations, on the other hand, are characterized by dispersal of microcomputers (or terminals of a centralized computer) throughout the firm. In contrast to the more centralized model, at satellite stations there is more interaction between attorneys, paralegals, and secretaries—primary work groups with a more fluid division of labor and more personalized relationships. Murphree found that technical control tends to be deemphasized with the satellite model.

It seems clear that computerized systems of office automation can serve to extend technical control—for instance, by monitoring productivity and counting errors. Garson, for instance, found extensive monitoring of airline reservations clerks, with such information as number of calls, percentage of bookings, dollar value of bookings per agent per hour, average length of calls, average number of seconds between calls, and percentage of time on line all routinely recorded.[61] Garson also found a high degree of standardization of the clerks' conversations. Even when workers work in apparently decentralized teams, a superficial, visible decentralization of the hardware (e.g., a computer terminal for every worker or every team) may mask an underlying centralization of control that is programmed into the software. Another more insidious form of control is subliminal messages that flash onto VDT screens, such as "You're not working as fast as the person next to you," or "Relax," or positive slogans about the employer and the workplace.[62]

Monitoring through computer systems has various effects on workers. Although it has been found to intensify stress,[63] it also appears that a certain computer mystique can offset dissatisfaction. Burris found that among the clerical workers she interviewed, "working with computers" was seen as inherently important work, and the fact that errors and productivity were monitored was taken as an indication of the importance of what they were doing.[64] The National Research Council discovered that attitudes such as "Everyone knows that computers make jobs better" can serve to offset dissatisfaction.[65] However, as Shaiken points out, centralized systems that stress technical control can increase worker dissatisfaction and backfire in terms of productivity, not only because of higher turnover rates but also due to worker strategies that foil the computerized monitoring process at the expense of customer service: hanging up on customers or minimizing time spent with customers in order to beat the computer clock.[66]

Some clerical and data processing jobs are now being exported, in much the same way that labor-intensive production jobs are. Gregory writes of this phenomenon:

> The "offshore office" provides yet another parallel to the manufacturing experience. . . . A certain amount of bulk information processing has been performed outside the country for some time. In the past, this work was shipped to and from offshore locations by plane, but the advent of satellite communications links makes the practice more attractive. One entrepreneur, George R. Simpson, chairman of New York-based Satellite Data Corporation, relays printed materials by satellite to Barbados, where the work is done by data entry clerks whose pay averages about $1.50 per hour. In Simpson's words, "We can do the work in Barbados for less than it costs in New York to pay for floor space. . . . The economics are so compelling," he told Business Week, "that a company could take a whole building in Hartford, Connecticut, and transfer the whole function to India or Pakistan."[67]

This trend toward exporting clerical work is a central reason why clerical work is predicted to decline as an employment sector in the United States.[68] "Underpaid and unorganized women in Jamaica, Korea, Singapore, Barbados, Mexico, and Haiti now perform much of the clerical work for New York City corporations."[69]

GENDER IMPLICATIONS

As bureaucracies are restructured into technocracies, the ensuing organizational changes have particular implications for women. Organizational gender barriers, paralleling divisions between skilled work and routine work, are not new and have accompanied various types of mechanization. Priscilla Murulo conducted an historical analysis of corporations in the late nineteenth and early twentieth century and their restructuring around office machines and Taylorism, concluding that "feminization and rationalization went hand in hand."[70] It is perhaps not surprising, then, that contemporary rationalization around computer systems is also gendered, that "automation and feminization are proceeding as twin and highly interrelated processes."[71] However, the restructuring of offices around computer systems is resulting in both opportunities and structural barriers for women, and represents both a continuation of and a departure from past experiences.

As middle-level positions and internal job ladders are eroded, the ensuing polarization into expert and nonexpert sectors appears to have paralleled and reinforced gender segregation and further circumscribed mobility prospects for nonexpert-sector women in organizations.[72] Early studies of computerization revealed that as many as 95% of upper-level technical jobs were filled by men.[73] More recent studies have found gender (and racial) segmentation to be still evident, although the disproportionality is less extreme.[74] Moreover, the jobs that are disproportionately filled by women are also more likely to be displaced by advanced technology.[75] Women have long been segregated into clerical and secretarial jobs with restricted mobility prospects, but in technocratic organizations this process is extended and structurally legitimated: The wider structural gap between expert and nonexpert sectors creates new types of credential barriers for nonexpert-sector women.

It appears that gender and race segregation are most pronounced in high-tech industries. Colclough and Tolbert found that women and racial minorities were concentrated in the nonexpert sector of high-tech firms, with white males dominating the expert sector: in 1978, 76% of electronics operatives were women, and 45% to 50% of the women operatives were third-world immigrants.[76] Conversely, they cite a recent study that found that 83% of professionals and technicians in semiconductor firms are white males.[77]

Colclough and Tolbert argue that there are three main reasons for this segregation: (1) gender stereotypes that attribute greater patience, eye-hand coordination, and manual dexterity to women (and perhaps to racial minorities as well);[78] (2) the fact that women and minorities constitute a flexible labor force, particularly important in high-tech industries, with periodic layoffs legitimated by the ideology that women are secondary wage earners supported by male breadwinners; and (3) the fact that white males are overrepresented in technical, scientific, and engineering disciplines (see chapter 5 for a fuller discussion of this last point).

Technocratic restructuring has occurred during a period of increasing equal employment opportunities for women and racial minorities, serving to thwart legal and social reforms at the organizational level. Noyelle discusses the EEO hiring and internal promotion decrees of the 1970s in the following way:

> A central principle behind these decrees was to eliminate discrimination by extending the benefits of internal labor markets to women and minority workers.
>
> Had everything else remained unchanged, the EEO challenge would probably have considerably weakened sex and race discrimination in the workplace . . . but at the same time that EEO policies were gaining speed, other forces came into play that began weakening the role of internal labor markets across a broad range of industries. Hence a basic dimension of EEO strategy—aggressive internal promotion of women and minority workers—was undermined. Some women and minority workers continued to advance to higher echelons, but their progress became increasingly dependent upon a different set of factors, involving educational credentials.[79]

If educational credentials are increasingly a prerequisite to upward mobility, then most nonexpert-sector women and minorities are likely to remain in low-level jobs. Murphree found that working-class women and minority women were more likely to work in the centralized, factory-like word processing centers, with men and more privileged women in the "satellite" centers.[80] Certainly some women have made major incursions into managerial, professional, and technical ranks, particularly in certain industries; for instance, during the 1970s women essentially doubled their numbers in the expert sector of the insurance industry.[81] It appears that women are not universally disadvantaged by recent organizational/technological change, but rather that new types of structural barriers have differential effects on women, depending on their class and race.

As personal secretaries become the exception rather than the norm, and word processing centers (centralized or decentralized) more prevalent, the social relations of production are correspondingly changed. Personalized patriarchal control (e.g., of a secretary to a male boss) is less apparent, although the subordination of the clerical sector to an expert sector that is predominantly male implies the ongoing significance of patriarchy: a more institutionalized, structural type of patriarchy, mediated by the technology and legitimated by the ideology of scientific knowledge and technical expertise. As Anne Machung puts it, "the seeming neutrality of the new word processing technology conceals the gender divide."[82]

Even expert-sector women face certain challenges and obstacles. Because they are more likely to work in relatively autonomous teams and task forces, interpersonal relations therefore become more salient, and the micropolitics of small-group interaction is heightened.[83] Leadership of such groups has been found to be derived from such sources as "information, expertise, connections, energy, creativity, and charisma,"[84] and given the predominance of men in expert sectors, imbalances of numbers and inequities of power imply that such micropolitics are likely to be gender laden. In the expert sector, with its majority of men and minority of women, one of the most salient aspects of the workplace culture is the emphasis on expertise and the need to prove oneself and one's expertise; "conspicuous expertise," a constant fear of "not being the expert," and the common response of doing a "snow job" to impress others have been found to be characteristic of expert-sector micropolitics.[85] Gender contributes to this insecurity over expertise in several ways. First, conspicuous expertise is more consistent with traditional male gender, giving men some advantages in the new types of micropoliticking. Masculinity has traditionally been associated with proving one's expertise, particularly in technical matters.

If technical expertise is more consistent with masculine self-identity, there is also the greater likelihood of *insecurity* among men in the expert sector: fear of being unmasculine if one's expertise is questioned. As Cynthia Cockburn points out, the competition among men to demonstrate technical competence is pronounced:

Male self-identity is won in a costly tussle with other men for status and prestige, and this applies in technological work no less than in other situations. Those men who seek their masculine identity in technological competence find themselves obliged to manoeuvre for position and negotiate their rank relative to other men. There are comparisons of competence: the *cognoscenti* versus the rest.[86]

Moreover, the threat from women in the expert sector is another source of insecurity; to be perceived as less competent than a woman would constitute a double threat to one's masculinity.

Given the emphasis on task forces in the expert sector, the micropolitics of small-group interaction, heightened by gender insecurity, is increasingly salient. Murphree, for instance, found that in the legal firms she studied, "life at the satellites . . . is not necessarily a bed of roses. . . . The primary-group nature [of these groups] may well encourage the same sexist and autocratic patrimony that characterized the former attorney-secretary dyads."[87] Natasha Josefowitz found that in small-group discussions, men often failed to hear women's comments, or attributed them to a male participant.[88] Similarly, Liz Gallese found that even women with elite credentials were not fully accepted into the upper echelons of corporate life.[89] It appears that one tactic that expert-sector men use to protect themselves from female competition and perceived threats to their expertise and masculinity is to deny expert-sector women recognition and visibility. As Judy Wacjman notes, one consequence of the traditional association of skill with masculinity is that skilled women are often not perceived as such.[90]

It also appears that the deemphasis on bureaucratic rules and norms in the expert sector, as well as the more relaxed and informal atmosphere of small-group interaction, encourages sexualization of the workplace in ways that emphasize gender. Cockburn, for instance, has documented how such informality, combined with the waning of traditional gender assumptions, has promoted a new type of masculinity.

> Women had identified this new type of male. . . . What distinguished him was an overt and confident machismo. Women everywhere made reference to the "cod-piece wearing jocks" of the policy unit, the "new men" of the advertising department. This masculinity does not share the woman's-place-is-in-the-home mentality of the old guard. These men expect to find women in the public sphere. Nominally at least they welcome women into this exciting new world because their presence adds sexual spice to the working day.[91]

In the context of competition and insecurity, as well as the general culture of what Jeff Hearn calls "hierarchic heterosexuality,"[92] this sexualization is laden with politics. The workplace culture is sexualized in such a way as to promote male solidarity, compensating for the competition over technical expertise (even though a good-humored male competitiveness over sexual expertise is also common). As Cockburn puts it: "Such sexualized discourse is a necessary part of the cementing activities with which senior men seek to bond men beneath them firmly into the fraternity, healing the contradictions of patriarchal and class structures that threaten to divide them."[93] Women tend to be disadvantaged by workplace sexuality, even consensual sexual relations, because of cultural and organizational stereotypes and double standards. They are in a no-win situation: If they attempt to participate as equals in sexual discourse and practice, they are perceived as dangerous *femme fatales*; if they resist or question sexualization they are labeled "iron maidens."[94]

Given the politicized context, sexuality always has the potential to become sexual harassment. Within the expert sector, where masculine insecurity over technical expertise and perceived threats from competent women are common, sexual harassment becomes one way in which "emergent and potentially powerful women . . . [can be] cut down to size."[95] Sexuality becomes a trump card of masculine privilege, a way of asserting power when other avenues fail. Given the imbalances of numbers and the isolation of women within the expert sector, resistance to such harassment is more difficult.

The dynamics of sexual harassment within the nonexpert sector are somewhat different. In the clerical sector, where women comprise the vast majority of the work force, sexual harassment is likely to come from male superiors. In such cases, the subordinate organizational position of women, corresponding to the general subordination of women in the culture, serves to make nonexpert-sector women vulnerable to sexual harassment. Indeed, the polarized and sex-segregated technocratic organization can be perceived as a fraternal group of expert-sector men with a virtual "harem" of nonexpert-sector women providing various types of support for male activities. Some men undoubtedly see sexual services as part of the fringe benefits of occupying privileged positions within the organization.

Despite these various types of gender discrimination, there is also the possibility that computerization may eventually reduce the importance of face-to-face interaction and attenuate the salience of ascribed characteristics. Zuboff, for instance, found that the computer conferencing system of the large pharmaceutical company she studied had begun to transform communication among the expert sector, with implications for women and minorities.[96] As one user of the system said,

> DIALOG lets me talk to other people as peers. . . . All messages have an equal chance because they all look alike. The only thing that sets them apart is their content. If you are a hunchback, a paraplegic, a woman, a black, fat, old, have two hundred warts on your face, or never take a bath, you still have the same chance. It strips away the halo effects from age, sex, or appearance.[97]

Of course, the fact that this person associates being a woman with social and physical handicaps may also be taken as an indication of the ongoing importance of sex discrimination. Whether computerized communication will become prevalent enough to undermine face-to-face interaction and such overt discrimination remains to be seen.

It appears, then, that women and racial minorities, particularly in the nonexpert sector, are often disadvantaged by the contemporary restructuring of bureaucratic organizations. Because they have tended to be concentrated in more routine work, they are vulnerable to technological displacement, part-time work, or temporary work.[98] When working with computerized systems of office automation, they are more likely to be monitored and controlled by the system. Finally, in organizations that favor team organization, they are sometimes discriminated against in the politics of small groups. As bureaucratic rules are relaxed and decentralized group authority is emphasized, women and minorities may become more vulnerable to personalized racism and sexism.[99]

Another recent organizational change with special implications for women is the experimentation with "telecommuting" from home: "the total or partial substitution of the daily commute by communication via a computer terminal."[100] Although home-use computer terminals are currently rather limited in their application, some see them as a viable future trend and a partial solution to the special problems of working mothers. One business writer, for instance, predicted that "portable terminals will be a special aid to homebound workers, such as mothers with small children."[101] In effect, home-based telecommuting represents one possible resolution of the central contradiction between capitalism and patriarchy; capitalism needs a low-level clerical work force, whereas patriarchy endorses a new version of the old idea that "woman's place is in the home."[102]

In a context of technocratic polarization, it appears that telecommuting is having divergent effects, based on occupational status and gender.[103] Professionals and managers, often allowed to work at home so that management can retain scarce talent, enjoy the flexibility and autonomy of working away from the office, using "electronic mail" to enhance communication and obviate the need for physical proximity to colleagues. The relatively autonomous working conditions of expert-sector workers makes working at home merely an extension of this autonomy.[104]

Clerical telecommuters, on the other hand, face a different situation.

> They are more likely to be paid piece rates, less likely to receive fringe benefits, and are sometimes required to pay rent for the equipment they need to do the work. . . . Supervision of off-site clerical workers is viewed as a problem by management. Companies deal with this problem by giving clerical telecommuters discrete tasks that have definite beginning and ending points and that can easily be measured and by paying piece rates.[105]

Clerical telecommuters are also sometimes monitored via the technology.[106] Telecommuting may be used to privatize workers, combat unionization, decrease worker benefits, obviate the need for child care, and increase the amount of housework allocated to women. Judith Gerson interviewed 106 home-based clerical workers and 260 office-based clerical workers and found that the home-based workers did more housework and child care than their office-based counterparts.[107] Margrethe Olson and Sophia Primps found that clerical telecommuters were isolated and experienced increased stress due to combined work and child-care pressures.[108]

Although there are also benefits to working at home, such as greater flexibility and a certain degree of autonomy, home telecommuting must be carefully analyzed in terms of its social implications, with the understanding that the form that telecommuting takes is political rather than technical in nature.

THE TRANSFORMATION OF MANAGEMENT

Managerial roles are also being transformed as bureaucracies are restructured into technocracies. Many middle-management jobs are being replaced by technological systems, and remaining managerial jobs are undergoing change in response to technological and social restructuring.[109]

Technological systems, sometimes worldwide in scope, both generate and require large amounts of information and data. Upper-level management may be inundated by large amounts of data, making it difficult to simultaneously understand the past, manage the present, and plan for the future. As technological systems increasingly require accurate data in order to function maximally, workers at all levels are constrained (although perhaps not all equally).

In order to effectively deal with an abundance of information, many firms are turning to computerized systems of management science and computer modeling of marketing decisions, resource allocation, and long-term planning.[110] Mathematical modelers now offer their services to public-sector managers as well, modeling social problems and suggesting solutions. These attempts to rationalize managerial decision making are worth examining in more detail.

The management scientist constructs a model by identifying important aspects of the problem, entering them into the system as variables, specifying a criterion, and then using computerized computation to determine the best policy decision. Planning decisions can be modeled by incorporating assumptions about the future into the specified variables. This process involves the substitution of decontextualized, abstract information for the more holistic, concrete, and experiental knowledge that has traditionally been the prerequisite of sound managerial decision making.[111]

In contrast to the more traditional way of training managers—slow upward mobility through the ranks, with accumulated expertise pertaining to all aspects of the company paralleling a growth in wisdom—the new context is that of formalized credentialing and upward mobility through interfirm mobility.[112] This context further promotes a reliance on abstract, formalized knowledge rather than concrete experience.

> What does the typical American manager bring with him when he changes companies? Not, unfortunately, much of the know-how he presumably acquired on the basis of concrete experience in his previous job. No two companies are exactly alike in personnel, problems, or philosophy. The manager's experiences in his old job must be translated into facts and general principles before they can be brought to bear in his new position . . . when holistic concrete experience is decomposed and transformed into rules, a great deal of its content is lost.[113]

The stage is set, therefore, for greater reliance on computer modeling and mathematical rationalization of management science.[114]

Dreyfus and Dreyfus argue that "the scientific combination of factors by means of formulas may well be one of the culprits behind our current business woes,"[115] since intuition and the "art" of management are being lost. Similarly, Zuboff interviewed a comptroller of a bank, who said of the shift to a sophisticated computerized information system, "People become more technical and sophisticated, but they have an inferior understanding of the banking business."[116] Another manager Zuboff interviewed said:

> If I didn't have the Overview System, I would walk around and talk to people more. I would make more phone calls and digress, like asking someone about their family. I would be more interested in what people were thinking about and what stresses they were under. When I managed . . . without it, I had a better feeling of the human dynamics. Now we have all the data, but we don't know why. The system can't give you the heartbeat of the plant; it puts you out of touch.[117]

Just as the subtle art of machining is lost in numerical control, computerized management may be similarly impoverished.[118] Moreover, just as with numerical control, valuable skills that are imperfectly approximated by computer systems are being eroded in the work force, leading to the possibility of a "skill crisis."

Deskilling (as well as displacement) is even more apparent at middle-management levels. As middle-level management decisions such as underwriting or loan approval are increasingly reduced to formulas and computer programs, with management trainees essentially only inputting data and responding to the system's questions, the middle manager's job is becoming confused as to purpose and function. Zuboff found that increasingly the question "What do we need middle managers for?" was being asked.[119] The danger, as middle-level positions continue to be eliminated, is that competence will be superficial, and eventually very few people will have enough understanding of the process to be able to adequately deal with atypical cases. Only if nonexpert-sector workers' jobs are informed can this danger be averted.

Computerized management systems can also be used as tools, however, enhancing and extending managerial judgment rather than substituting for it, in much the same way that numerical control can be used by skilled machinists to extend and perfect their skill. Some problems can be modeled in order to enhance and perfect decision making, and modeling may be particularly helpful in dealing with problems without precedent—such as understanding the likelihood of nuclear winter. Another related change in managerial roles has to do with the increased salience of technical experts in both the private and the public sectors, and the increased necessity for managers to successfully interact with these experts and integrate technical expertise into decision making. As we saw in chapter 3, in some cases conflicts between managers and technical experts have arisen, focusing on questions of managerial versus technical authority and leading to the integration of technical and managerial roles. In some instances, managers' lack of technical knowledge has undermined their authority with technical workers, as traditional rank authority becomes less important than the use of "influence."[120] "Conspicuous expertise," credentialism, and other types of cultural capital have become increasingly important political currencies in technocratic organizations.[121]

The public sector is an advantageous setting in which to explore the politics of technical expertise and its effect on decision making. Three recent books have analyzed the ways in which technical experts have changed the shape of public sector decision making: Elliot Feldman and Jerome Milch's *Technocracy versus Democracy*, Serge Taylor's *Making Bureaucracies Think*, and Jeffrey Straussman's *The Limits of Technocratic Politics*.[122] Feldman and Milch compared international airport construction and land-use policy in the United States, Canada, Italy, Great Britain, and France; Taylor studied the U.S. Environmental Protection Agency; and Straussman analyzed the growing role of economists, social indicators experts, and futures research experts in U.S. government policy making.

Feldman and Milch contend that in the 1970s, public-sector bureaucracies were encouraged to adopt more businesslike techniques in order to make them more cost-efficient and effective. A major aspect of this shift has been the emergence of governmental technocracies.

> Bureaucracies populated by careerists began giving way to technocracies populated by specialists. . . . where governments could not staff themselves with experts they hired outside consultants. The basic direction, throughout the advanced industrial world, has been to entrust greater decision-making authority to experts who are armed with the techniques of rational planning, or officials who are urged to acquire and use them."[123]

Feldman and Milch conclude from their international case studies that technocratic experts have fundamentally altered the shape of public-sector decision making, although not necessarily for the better. The incorporation of technical expertise has altered the terms of the debate and made policy making more difficult, as experts often give conflicting advice. The ideology of rational planning says that there is "one best solution" to any conflict of interest, but in practice experts reach different conclusions and advocate different solutions. As a result, "technical choices never depend exclusively on merit; they are always political."[124]

Feldman and Milch found that business motives are highly salient behind the scenes, with a corresponding narrowness of goals and an inattention to the public interest. They therefore conclude that business goals, however rationalized by expert advice, are inappropriate for public-sector decision making, leading to irrational policies such as acrimonious land disputes when new airports are built, culminating in the highly contested land being unused for decades. Technocracy's erosion of democracy can backfire, implying that future rationalization must be of a political nature.

> The most important lesson that civil aviation can draw from the experiences of these five countries is the importance of establishing new and healthier relationships with the public. Aviation enthusiasts have demanded public support for unlimited growth on the basis of a technological mystique and the knowledge of experts. The controversies of the 1970's demonstrated that there is no monopoly on expertise because experts frequently disagree. Support for development will have to be won in the future more through public debate and persuasion than through technological mystique.[125]

Taylor reaches similar conclusions from his analysis of the Environmental Protection Agency. As the agency's decision making has become more technical and scientific in nature, the result has been the attempt to rationalize the bureaucracy by importing scientific norms and procedures. To operate effectively in a highly technical and changing environment, organizations must exhibit the capacity to learn. Taylor concludes that although the EPA has attempted to emulate scientific models of learning and rational decision making, these attempts have been of limited success.

The main reason for the difficulty in grafting scientific procedures onto public bureaucracies is the absence of public consensus and the inability of science to point towards one best solution. Science can provide a certain measure of *instrumental* rationality by illuminating the relative utility of various means; however, more *substantive* rationality of goals and ends will remain controversial. Taylor concludes that a complex combination of science and politics emerges:

> A science-like process in politics must base its substantive rules on more than empirical criteria. These rules must appeal to general social values as well as empirical evidence. . . . Where science leaves off, politics must fill in. . . . In other words, science provides one form of consensus, an unforced consensus based on logic and empirical evidence. An equally workable consensus is provided by a society with homogeneous conventional beliefs. When neither scientific nor conventional consensus is strong enough, politics is called upon to fill the spaces left by incomplete knowledge. Where a wide and deep consensus does not exist, politics is left with the hard problems. . . . science cannot provide unequivocal answers to all important questions.[126]

Straussman also concludes that the increased salience of technical experts within the U.S. government has changed rather than diminished the role of politics in public sector decision making. "Politics becomes a matter of choice between a limited range of options,"[127] with the options determined by both technical *and* political factors. A new type of "technopolitics" appears, as experts serve both guidance and legitimation functions. The assumptions made in economic and futures modeling incorporate political decisions, but the latter are typically not seen as such. Also, final reports get "sanitized" according to political motives but are taken as entirely technical and rational in nature. The scope and nature of technical expertise is therefore narrowed in accordance with political interests: "Experts are on tap but not on top."[128]

Feldman and Milch, Taylor, and Straussman all agree that public sector management has been transformed by the increased salience of technical experts and expertise within public bureaucracies. Like computerized aids to management, technical experts are used and managed for political ends. The nature of politics has therefore been transformed, but not rendered obsolete, by the increased salience of technical expertise and technical experts in government. All three books see the legitimation of political decisions on technical grounds as potentially dangerous to democracy, and the erosion of democracy as counterproductive for public bureaucracies in the long run.

In a study of an even more politicized public bureaucracy, Greg Bischak studied the regulation of the nuclear industry and the Advisory

Committee on Reactor Safeguards (ACRS).[129] He found that the engineers within the ACRS were able to exert influence in the direction of safety regulation, but *only* after substantial public pressure was already building outside the industry.

> The environmental and safe energy movements, together with progressive elements of the labor movement, created a context in which some segments of the regulatory technocracy could dissent from the bureaucratic dictates of the Atomic Energy Commission and later the Nuclear Regulatory Commission. By and large, though, the bureaucratic hierarchical decision-making process tended to overrule what the progressive technocracy suggested was the safe and scientific approach. Only when there was enough external political pressure applied from outside the process was it possible for the progressive regulatory technocracy to maneuver within the regulatory process to press for change. Once the external political pressures waned under Reagan, the technocracy had less room to maneuver. . . . scientific defections from a technocracy have much to do with the broader political culture which creates a context in which technocratic dissidents feel they can effect a change.[130]

CONCLUSION

When one examines recent conceptualizations of changes in bureaucracy in conjunction with recent empirical studies, what is most striking is the fact that most theorists focus on the expert sector in formulating their theories. As in the popular press, the more advantageous working conditions of the expert sector are taken to be generally characteristic. Resulting conceptualizations (e.g., "adhocracy," "rational-productivity bureaucracy," "integrative organization," "technobureaucracy") therefore ignore both the nonexpert sector and the overall nature of work organizations.

The most common empirical finding among these studies of changing bureaucracy is that of polarization. Expert and nonexpert sectors not only are clearly defined, but are also increasingly separate from one another, as internal job ladders are eroded. This transformation of the division of labor is due to the systematic reorganization around technological systems, which substitute internal, technical complexity for the organizational complexity characteristic of bureaucracy. Many skilled, semiskilled, and supervisory functions are increasingly performed by computer systems, leading to a bifurcation between a managerial/technical sector and a less skilled sector. The most advanced technological systems and the most complete automation lead to the most pronounced polarization, with the nonexpert sector performing very low-level data-entry and data manipulation functions. However, even in industries that are resistant to such high levels of computer automation, such as insurance, the option of emphasizing skilled clerical workers is nonetheless exercised within the context of an overall polarization of the firm into expert and nonexpert sectors.

The polarization of structure is paralleled by a polarization of working conditions. At expert levels, much of the rigidity and centralized authority of conventional bureaucracies is relaxed, leading to more flexible teams and task forces; the privileges and autonomy of expert-sector workers are similar to those that professionals working within bureaucracies have typically enjoyed. Traditional managerial authority is deemphasized in favor of influence and technical expertise.

At nonexpert levels, however, a wider range of options is available. In some firms, a "mechanistic" type of bureaucratic structure at the nonexpert level coexists with a more "organic" model at upper levels. Workers may be monitored and paced by the computerized system, performing fragmented tasks in an isolating and stressful environment.

In other firms, organizational innovation in the nonexpert sector has resulted in experimentation with teams, job rotation, satellite stations, and task reintegration. Even for workers in such firms, however, working conditions are often inferior to those of the expert sector. External status differences and disparities in knowledge may thwart attempts at egalitarian working relationships. In some cases, only very superficial attempts at democratization are made. The emphasis on decentralized authority appears to make some workers more vulnerable to personalized discrimination based on sex, class, or race; conversely, some argue that the emphasis on computerized data and communication may promote greater objectivity and less face-to-face discrimination. Visible decentralization of the computer hardware (e.g., computer terminals for every satellite station) may mask underlying managerial control programmed into the software or may promote a more genuine decentralization of power.

In looking at the overall organizational structure, one finds both some general tendencies (e.g., polarization, increased salience of technical expertise and technical experts, skill restructuring) and a wide range of diverse options, particularly for the nonexpert sector. Technological systems can be implemented so as to increase task fragmentation or to promote task reintegration around skilled clerical workers or multiskilled teams. The job of the skilled clerical worker can be

viewed as an upskilling of the previous clerical job or as a deskilling of
the previously more skilled job (e.g., underwriting). Clerical workers
can be upskilled but simultaneously monitored by the technological
system. Computer systems can be designed to intensify centralization
and technical control or to promote decentralization.

Given the wide range of options, politics becomes more, rather
than less, important in technocratic organizations. Recent studies of
public-sector policy making reveal the ongoing importance of politics,
as well as the ways in which political maneuvering can be masked by
allegedly technical decision making. Analyses of the rapid changes in
private-sector bureaucracies indicate that political motivation may be
similarly veiled; future rationalization of work organizations must be of
a political nature.

Chapter 5

THE TRANSFORMATION
OF PROFESSIONALISM

*Professional education trains recruits in specialized knowl-
edge and skills but also turns them away from an impracti-
cal concern with moral ideals toward a "realistic" and appro-
priately pragmatic focus on the technical expertise of the
profession.*
—*Charles Derber,* Professionals as Workers

An understanding of recent changes in professionalism is central
to an understanding of technocracy, for those who exercise expert
power are usually categorized as professionals, and knowledge and
technical expertise are translated into societal power to a significant
degree through professional strategies.[1] Professionals are the "agents of
formal knowledge,"[2] and professionalism has changed as formal
knowledge has changed. As Magali Larson points out, analyses of the
professions raise broader and more fundamental questions concerning
the social production and certification of expert knowledge, as well as
the implications of these for democracy.[3] In order to understand contemporary professionalism, as well as the larger social questions that it
raises, we must examine recent conceptualizations of changes in professionalism and recent empirical studies of professional work.

Preindustrial definitions of professional work, such as medicine,
theology, and law, centered around its high status and the aristocratic
nature of its practitioners.[4] It was during industrialization, however,
that the classic model of professionalism emerged. Classical profes-

sionalism emphasizes specialized expertise or skill, increasingly derived from formal training and certified by credentialism, which differentiates the professional from other workers. Moreover, professional expertise allegedly combines formal and tacit knowledge, intellectual and practical knowing, into a uniquely valuable synthesis yielding a mysterious and charismatic control over unpredictable and stressful situations.[5] It is this unique and individual quality of professional expertise that is the basis for professionals' claims to autonomy and self-regulation of their work and training.[6] Moreover, classic professionalism emphasizes a service ideal and ethical standards of altruism, both serving as additional justification for high social status and minimal societal regulation of professional conduct.

However, with recent socioeconomic changes, such as the increased size and scope of the public sector, the development of complex technological systems, and the increased salience of technical expertise and specialized knowledge, professional work has changed accordingly. Have professionals become more integral to the capitalist economy and therefore more powerful, or have traditional bases of professional power been undermined by recent socioeconomic developments? How have professionals responded to these changes?

CONCEPTUALIZATIONS OF CHANGING PROFESSIONALISM

Analysts of professionalism have generally agreed that the nature of professionalism has changed in recent years, as the ideal type of the autonomous, self-employed professional has become the exception rather than the rule.[7] However, conceptualizations of these changes have varied, with little consensus as to either predeterminants or implications.

One school of thought argues that deprofessionalization is occurring.[8] Marie Haug, who has been the main spokesperson for this perspective, argues that traditional definitions of professional work have become obsolete due to major changes in the organization of professional work and the relationship between professionals and clients. One source of change is rising levels of public education and an alleged demystification of professional knowledge and power. Moreover, Haug argues that a growing societal consciousness of the need for professional accountability and the dangers of professional malpractice have altered the professional/client relationship and undermined professional power.[9]

Another primary source of deprofessionalization, according to Haug, is technological change. She argues that computerization has eroded professional power by making the knowledge once monopolized by professionals more specialized, codifiable, and generally available to a wider audience.

> Scientific professional knowledge can be "codified"; it can be broken into bits, stored in a computer memory, and recalled as needed. No longer need it be preserved in the professional's head. . . . And what is put into the electronic machine can be extracted by anyone who knows the output procedures. Command over the stored knowledge is not because one knows it but because one knows *how to get it*.[10]

If professional knowledge must be neither overly specialized nor overly broad and common, but rather a combination of tacit and esoteric knowledge, then Haug is arguing that computerized knowledge is both overly specialized and overly available and hence resistant to professional monopolization.[11] In fact, she contends that paraprofessionals and technicians (for instance, in medicine and law) will increasingly take over professional functions with the aid of computerized technology.

A different but not unrelated approach to understanding recent changes in professional work is that of the *proletarianization* of the professions.[12] Adopting a more Marxist perspective, one that analyzes the changing political economy of capitalism and the implications of such changes for the labor process, these analysts make analogies between classic proletarianization of manual labor and recent developments in professional work.

Larson contends that the recent trend toward professional employment in large bureaucratic organizations, combined with corporate pressures to achieve profit maximization and public-sector fiscal crises, has led to professional proletarianization as evidenced by three major tendencies: (1) a more narrow and rigid division of labor, (2) an intensification of professional labor (heavier caseloads, increased volume of work), and (3) the routinization and standardization of high-level tasks, and the corresponding possibility of delegating such tasks to less educated workers or computers.[13] The implication of these trends is that "professional status . . . no longer insures the incumbent against the predominant relations of production in our society."[14]

According to Larson's analysis, these changes in professionalism imply a "technobureaucratic professionalism" for professionals

employed in large bureaucracies. One implication of technobureaucratic professionalism is the transformation of the traditional client orientation. Instead of traditional client service, Larson argues, client
manipulation, client indifference, client hostility, or, in some politicized
instances, political coalitions between professionals and clients for organizational change may result.[15]

Larson concludes that the technobureaucratic professions retain
only one characteristic of classic professionalism: specialized skills. As
traditional professionalism is transformed, the ideology of professionalism (craftsmanship, service, noblesse oblige) is emphasized as an
"internalized mechanism for the control of the subordinate expert"[16]
and in order to give a sense of prestige to proletarianized professionals
who might otherwise be discontented with their actual working conditions. Other managerial strategies used to keep professionals satisfied
include organizational segregation, relative autonomy, and the retention
of considerable discretion over one's work (as distinct from organizational power, which Larson argues is retained by management).

Charles Derber summarizes the proletarianization perspective
somewhat differently.[17] He sees the development of advanced computer technology to assist professional work as detrimental to professionalism primarily because it makes it impossible for professionals to
own and control their means of production, just as the technological
developments of the industrial revolution undermined craft work. This,
in turn, implies that professionals become dependent upon large institutions for their economic survival, and the distinct possibility that
these institutions will perceive the need to control both the technology
and the professionals who work with it. "Professional autonomy is
reduced to technical discretion."[18] Client loads increase, further expanding the need for a centralized administrative apparatus to regulate professional work. Derber concludes:

> As "knowledge" and service markets have expanded and become highly
> profitable in the evolution of the twentieth-century postindustrial econ-
> omy, and profit margins in the traditional manufacturing sector of the
> industrial economy decline, capital pours into expanding medical, sci-
> entific, high-technology, and service sectors. The meaning of this mas-
> sive shift in the flow of capital is that service and professional areas tra-
> ditionally organized according to "petit bourgeois," craft, or cottage
> industry models of small production become reorganized along the lines
> of the dominant centralized corporate economy. . . . The "proletarianiza-
> tion of the professional" is thus an outcome of the capitalization of men-

tal labor and human services, which has swallowed up the last remaining realms of "free" labor in the economy and signifies a decisive consolidation of corporate and state power.[19]

Derber's own theory of changing professionalism expands upon Larson's idea of the relative autonomy of contemporary professional work. He contends that professionals are *technically* quite autonomous but *ideologically* subordinate to capitalism. Professionals are increasingly required to work closely with paraprofessionals and with advanced technology, but they are not deskilled or fundamentally restricted in their autonomy. Ideologically, however, professionals are required to accept the goals and interests of their employers, in contrast to traditional professional ends (maximizing their own status, client service, advancement of knowledge). The result is a managerial strategy of relative autonomy "that does not threaten them with deskilling or loss of their technical knowledge and autonomy, but nonetheless effectively subordinates them to the imperatives of capitalist production."[20]

In his more recent work, Derber (with William Schwartz and Yale Magrass) argues that the new class of professionals and experts has gained in power, vis-à-vis the capitalist class.[21] They see this growing power of experts as related to the periodic crises of capitalism.

> With each capitalist crisis, the new class gets its foot a little farther in the door to power. Each great cycle leaves behind a new planning apparatus—within both the corporation and the state itself—employing and depending on experts more and more. As the new global economy puts severe strains on capitalist systems, new-class Democrats in the United States may form the basis of an empowered professional class cadre in government. As in France and Japan, a permanent caste of logocrats may emerge: engineers, lawyers, accountants, economists, scientists, and planners who populate key policy agencies and become the mandarins of the new era.[22]

They contend that the inherent contradictions of the capitalist system are gradually resulting in its transformation into a new type of "mandarin capitalism": not the precise change that Marx foresaw, but a significant change nonetheless.

Other analysts have interpreted the changes in professionalization as less fundamental and as related to a more Weberian bureaucratization or rationalization of the professions. Gloria Engel and Richard

Hall, for instance, argue that although the professions are being restructured and "industrialized" in ways that parallel the industrialization of craft labor, these changes do not threaten but actually may promote, a "more faithful adherence to professionalism."[23] Professionals who are employed in bureaucratic organizations are freer from economic pressures, and Engel and Hall contend that they are therefore more likely to focus on service and altruism. Moreover, as more individualistic service gives way to a more hierarchical team approach, and as professional knowledge becomes more widely available, the opportunities for both peer and public evaluation are enhanced. The professional/client relationship may be less private and personalized than in the past, but this may actually improve the quality of service and protect against possible abuse of the relationship.

Raymond Murphy also contends that the professions are being bureaucratized and rationalized in ways that transcend capitalism, in that similar developments are found in state socialist societies.[24] For Murphy, the proletarianization-of-professions literature has been a detrimental detour that has led to a dead end. The formal rationalization of both capitalist and state-socialist societies has created new bases of power for professionals, based on their skills and credentials, even while the same forces have resulted in certain bureaucratic constraints on professional labor.

In a similar vein, Eliot Freidson argues that conceptualizations such as deprofessionalization or proletarianization are exaggerated, in that professionals continue to enjoy high status, prestige, and occupational power.[25] Computerization may not adversely affect professionals, since "it is the members of each profession who determine what is to be stored and how it is to be done, and who are equipped to interpret and employ what is retrieved effectively."[26] He contends that professionals are not being bureaucratized any more than bureaucracies are being professionalized—that the empirical literature demonstrates many examples of "hybrid forms of organization that deviate from the bureaucratic model in order to accommodate their professionals."[27] Freidson concludes that the professions are being "formalized": that although the control of professional work is still largely in the hands of professionals themselves, control mechanisms and internal stratification of professionals have become more formalized. He sees the institutionalization within professional organizations of an administrative elite, a knowledge elite, and the rank-and-file professionals, and a possible erosion of professional communities as a result.

Andrew Abbott takes a more systemic approach to understanding the contemporary rationalization of the professions.[28] For Abbott, the professions operate within an interdependent system, with various types of jurisdiction. Moreover, the jurisdictional boundaries are in constant dispute, leading to an evolutionary dynamic of the entire system of professions. In recent years, technological and organizational changes have led to the creation of new professions, the demise of others, and intensified jurisdictional battles among many professions.

More specifically, Abbott argues that professional work consists of three main components: diagnosis, inference, and treatment, with inference being the professional skill needed to effectively link diagnosis and treatment. For Abbott, either too little or too much use of inference weakens a profession's claim to jurisdiction and societal power: Too little makes the diagnosis and treatment sequence too routine to warrant professional status, and too much makes the work seem insufficiently legitimated by formalized knowledge. In recent years, computerization in the context of capitalism has promoted what Abbott calls the "commodification of knowledge" and a corresponding routinization and degradation of some professions.[29] Moreover, expert diagnostic systems are becoming effective competitors with professionals in such fields as general medical diagnosis, mass spectroscopy, tax law, air traffic control, electrical engineering, and so on.[30] Expert systems have the potential to transform the work and status of even the most elite professions; as Abbott puts it, "expert systems promise commodification well up the abstraction hierarchy."[31]

Abbott's analysis has affinities with those of both Marx and Weber. In some ways he sees professionals as a new class, possessing a new type of knowledge capital and yet forced to work for wages. However, this new class is lacking in interprofessional solidarity and is divided by the constant jurisdictional struggles among the professions. Abbott argues that expertise can be institutionalized in people (professionals), things (commodities such as computers), or organizations (an emphasis on bureaucratic rules and efficiency). Currently, Abbott sees professionals holding their own against the competing pressure of commodification and organizational control, but also losing ground.

In order to be in a better position to evaluate these divergent analyses of contemporary changes in professionalism, we turn now to recent empirical studies of different kinds of professional work.

MEDICAL PROFESSIONS

The medical professions, often taken by sociologists as paradigmatic of professional work, have undergone substantial structural
change in recent years. Such factors as growth in the number of for-
profit hospitals and clinics, changes in insurance programs, changes in
client attitudes and health care preferences, shifts from inpatient to
ambulatory care, and technological changes have been widely influential. The health care industry has witnessed above-average employment growth during the 1970s and 1980s, and this growth is predicted
to continue through the 1990s.[32] However, it is likely that the restructuring of the health care industry will continue, making future growth
patterns unpredictable. Although hospital employment grew by 50%
from 1970 to 1982, it declined slightly in 1984 and 1985, and projections
for the next decade predict a decline of up to 20%.[33] To better understand this restructuring, one must look at specific health care occupations and how they have changed in recent years.

Allied health workers, which comprised two-thirds of all health
workers in 1970, were projected to constitute at least three-quarters by
1990.[34] Some have predicted a trend toward minimal physician/patient
contact, with paraprofessionals, technicians, and medical technology
assuming such functions as medical history taking, physical examinations, and diagnostic procedures: a "postphysician era," in which paraprofessionals will monitor computerized examinations, testing, diagnosis, and medication.[35] In some settings, paraprofessionals and nurses,
with the aid of advanced technology, have taken over most diagnostic
patient care; for instance, "telemedicine," bidirectional cable television
systems, allows some community clinics to function without on-site
physicians.[36]

The context of physician care has also changed. Physicians are
less likely to be in private practice and more likely to specialize; in 1932,
more than three-quarters of all physicians were in general practice, but
by 1973 this figure had dropped to 14%, with projections of 6.4% by
the 1990s.[37] Increasingly, physicians work in "health care systems,"
dominated by sophisticated technology. A visit to the doctor becomes
an encounter with medical technology and paraprofessionals.[38]

However, some argue that to speak of the "deprofessionalization"
or "proletarianization" of physicians is an unwarranted exaggeration, as
physicians appear to have retained considerable occupational power
and workplace autonomy.[39] Advanced technology may actually be

enhancing, rather than undermining, the work of the physician.
Various computer systems have been designed to aid medical
diagnosis. These include MYCIN, a program to analyze the results of
blood tests and suggest blood disorder diagnoses; INTERNIST-I, a program for diagnosis in internal medicine; PUFF, an expert system for
the diagnosis of lung disease; and RECONSIDER, an interactive ency-
clopedia designed to provide a comprehensive list of possible diseases
when given the patient's symptoms.[40] The success rates of these programs have been variable, and interpretations have been diverse.[41]
McKinlay argues that "computers can perform nearly all aspects of
diagnosis, treatment, and therapy at a level of reliability higher than
that of even the most highly-trained and up-to-date physician."[42] Conversely, Dreyfus and Dreyfus conclude from their analysis that although
physicians (and patients) may find it comforting to believe that their
work is rule governed and derived solely from calculative rationality,
given the anxiety associated with medical decision making, the work of
the physician involves as much intuition as rational knowledge.

> Doctors are tempted to rationalize their intuitive decisions not only to
> justify them to themselves and their peers, but also in order to explain
> them to their patients. . . . [but] the doctor can never factually explain his
> innermost feelings about the preferred therapy based on a lifetime of
> experiences with similar cases. Every case is unique, so statistics about
> likelihoods of outcomes of various possible treatments based on all pre-
> vious cases are of little value. . . . In reality, a patient is viewed by the
> experienced doctor as a unique case and treated on the basis of intuitively
> perceived similarity with situations previously encountered.[43]

The general consensus appears to be that the various diagnostic
computer programs are tools for the physician to use rather than
replacements for the physician's expertise. The National Research Council concludes: "In all of this software, the physician has access to the
data and the rules that the software uses in ways that allow the physician to understand and challenge its assumptions and conclusions."[44] It
appears that tacit knowledge has not been rendered obsolete by recent
medical technology, but rather that such technology may serve to augment and legitimate professional judgment.

Questions about changes in the physician/patient relationship
are more difficult to answer, mostly because this relationship is so variable. Class differences, for instance, are widely apparent, with patients
of higher socioeconomic status more likely to have more personalized

and communicative interactions with their physician.[45]

Overall, there is some evidence of deteriorating physician/patient relationships. Terry Mizrahi, in a longitudinal qualitative analysis of professional socialization of interns and residents in the field of internal medicine in a large university medical center, found a pronounced antipatient orientation.[46] Patients were seen as creating work overload, including an overload of supervisory responsibilities, and there was a tendency to attempt to "get rid of patients" who could be referred elsewhere. She also found that the interns and residents objectified patients, ridiculed them, and at times wished they would die. One doctor, for instance, said:

> [When a patient dies], you're just interested in your own survival. Sometimes it's hard to relate to people who die, and it's a relief that they die, not only because they're so sick and what you put them through is often worse than death, but the fact that you don't have to do that anymore. It saves time and energy, so you think in terms of the overall benefit.[47]

Mizrahi concludes that "regardless of whether patients were physically harmed or neglected as a result of black humor or fantasized destruction, the result was a dehumanizing climate inconducive to a harmonious doctor/patient relationship."[48] Clearly the context of work overload is relevant to this situation, but she also found that physicians' attitudes were correlated with the race, class, and sex of the patient; patients of higher-status groups were often seen as more "interesting" and less self-destructive cases, more worthy of attention.

Another dimension of the transformation of medical training concerns increasing technological dependence and a trained incapacity to practice low-tech or independent health care. Low-tech medicine has been found to be held in low esteem by both patients and physicians, with good medicine increasingly defined as maximal use of the latest technology.[49] Similarly, based on diverse studies and polls, Derber concludes that first-year students of medicine, dentistry, and law are resocialized from a "moral" orientation to a "technocratic" perspective, learning "to perceive technical knowledge and competence as being of far greater importance than long-term ethical or social concerns"; "patient-centered idealism" gives way to "technique-centered pragmatism" and "realism."[50]

Yet another way in which medical practice is changing is that chronic diseases and corresponding medical technology designed to manage them are becoming more prevalent. One recent study, based on four years of observation and interviews at seven metropolitan hospitals, found specialization and the proliferation of paraprofessionals and technicians as one result of this trend, and fragmentation and depersonalization of care as another.[51] Long-term patients tend to become disillusioned with the medical profession.

Advanced medical technology is also related to the increased salience and transformed working conditions of health care technicians and paraprofessionals within the health care system. The social/technological changes in X-ray departments of British hospitals were studied by Cockburn, particularly the increased prevalence of CT (computer tomography) scanners. A prime exemplar of increasingly capital-intensive health care,[52] CT scanners have become widespread and routinely used in recent years. Unlike other technological changes, however, CT scanners represent a *supplementary* technology; more conventional X-ray technology (and other diagnostic techniques) are still used.

CT technicians are part of a complex hierarchy of medical personnel. They are paraprofessionals, subordinate not only to doctors, but also to the radiologists in charge of the scanner's clinical use. Yet another group, the medical physics department, is in charge of all equipment involving radiation. Cockburn found that these various groups of personnel are characterized by tensions and politicking centered around jurisdictional power. Physicists have gained power and prominence in hospitals, at times competing with doctors. Class and gender politics were found to be salient as well, with radiographers being disproportionately female and poorly paid relative to the doctors and physicists, who were predominantly male.[53]

Cockburn found that despite the relatively extensive training required for radiographers, technological change has increasingly been in the direction of more user-friendly scanners, implying a deskilling trend. Unlike conventional X-ray technicians, who exercise more judgment and discretion with regard to the positioning of the patient, the best views for a given diagnosis, selection of the best film and exposures, and so on, CT radiographers have more of a "push-button job."[54] Although the radiographer must know how to communicate with the scanner via a computer terminal, there is relatively little discretion, judgment, or knowledge of CT technology required for the job. In fact, the job is rather routine. The fact that CT radiographers often perform traditional X-ray work as well makes the deskilling less onerous, and a survey of radiographers in sixty British hospitals found that most would not want to work on CT scanners full-time.[55] Another aspect of the CT radiographer's work is the intensification of the work and the corresponding deterioration of their relationship with the patients. Due to the high cost of CT technology, optimal utilization implies large numbers of patients. In one hospital, Cockburn found that the patient load in the X-ray department (including one CT scanner and four conventional X-ray rooms) was one hundred patients a day, more than double the figure of four years ago, with no increase in staff since that time.[56] There is less and less time for interaction with patients.

Nursing is another health care occupation that has changed in ways that parallel the broader social and technological changes in the health care system. Computerized technology, increasingly linked to other health care technology, has begun to transform traditional nursing practice, at times leading to stress and/or resistance.[57]

One technological development that has the potential to radically change nursing is automated patient monitoring. Currently used in different forms and to variable extents in different settings, and salient particularly in intensive care units, automated monitoring has now been technologically perfected so that more extensive applications are possible. The National Research Council concludes:

> With the further development of the technology of representation of physiological processes and the reduction of noisy signals, technical experts saw total automation as a possibility. It has been suggested that fully automated systems could lead to a reduction in nursing staff time as well as in the medical/technical knowledge necessary on the part of nurses. Automated monitoring, combined with other technology, could result in continuous recording over extended periods, increased automatic storage and analysis of records, and automatic generation of suggestions for treatment and automatic administration of medication.[58]

At present, social preferences (of doctors, nurses, and patients) for traditional personalized nursing care has mitigated the technological trend toward automated patient monitoring, but future choices may be different, particularly given the shortage of nurses.

Finally, the health care system is changing in that it is becoming more integrated, more centralized, and more dominated by a corporate ethos.[59] The proliferation of for-profit multihospital systems and related health-services businesses, such as laboratory services and minor emergency centers, has led to dramatic changes that are projected to continue. In the future, the central issue for health care professionals may be less the bureaucratic infringements on their autonomy and more the tension between capitalist imperatives and traditional norms of professional practice. The increasingly capital intensive nature of medical technology, as well as the trend toward outside control of hospitals and other health care organizations, both imply increasing professional vulnerability to capitalist control.

THE LEGAL PROFESSION

Like the medical profession, the legal profession has undergone dramatic change in the post-World War II period. Bureaucratization, advanced technology, and an increasingly corporate ethos have transformed the work of professionals and nonprofessionals alike.

One of the first to explore this changing world was E. O. Smigel.[60] Given the increased prevalence of large, bureaucratic legal firms, he asked whether and how such contexts were transforming the nature of the legal profession. Smigel found that large corporate law firms were increasingly "spokesmen for big business"[61] and that lawyers trained in such firms received a "skewed view of the law" and a "trained incapacity" due to specialization.[62] He also found that lawyers in large corporate firms tended to become social conformists due to a narrow socialization.

However, Smigel concludes that the impact of bureaucratization on the legal profession was reduced by the fact that corporate firms recognized the need for professional autonomy and encouraged it. It was primarily professional norms of conduct that constrained the lawyers rather than managerial authority. The legal profession was, however, changing, as formal training programs replaced informal apprenticeship models and as stratification of the profession increasingly implied that "the wealthy and the powerful have access to the best attorneys."[63]

In the more than a quarter of a century since Smigel wrote, the legal profession has continued to change, leading some analysts to more negative conclusions, such as legal deprofessionalization.[64] Haug points to computerization of tasks formerly performed by legal professionals, such as the drafting of state legislation, screening of prospective jurors, and research of prior cases in case preparation.[65] She even sees the future possibility of using computers to actually plead and decide cases, replacing lawyers and judges in the courtroom.

In a more sustained analysis, Robert Rothman points to six reasons why the legal profession is becoming deprofessionalized.[66] First, the "competence gap" is narrowing, as laypersons are becoming increasingly knowledgeable about the law and taking a more active and critical role as clients. Second, professional knowledge is becoming routinized, not only by computerized technology, but also by the proliferation of do-it-yourself legal manuals. Some legal documents (e.g., wills, contracts) can now be computer generated when case-specific data is entered; another computer program predicted zoning appeals decisions with 99% accuracy.[67] Third, the legal profession is becoming highly specialized, not only according to type of law, but also according to type of clientele, type of practice (solo vs. firm, private vs. public) and prestige within the profession. Rothman argues that these internal divisions have reduced professional solidarity and that, because solidarity is important to the maintenance of professional status, that the latter may also be undermined. He sees an additional obstacle to solidarity in the increasing heterogeneity (gender, race, class) of the profession.

Fourth, the legal profession has become more consumerist and is increasingly viewed as merely one type of market exchange, making clients more likely to scrutinize legal services and less likely to accept professional self-regulation of legal practice. Fifth, the legal profession is increasingly suffering from the encroachment of other professionals, such as accountants, bankers, tax consultants, and realtors, undermining their monopoly of legal practice. Finally, Rothman argues that the employment of lawyers in bureaucratic organizations has undermined professional autonomy and discretion, due to standardization of procedures and centralized decision making (e.g., decisions regarding the disposition of major cases).

For all of these reasons, Rothman concludes that deprofessionalization of the legal profession is occurring, even in a context of an increasingly litigious society with large numbers of lawyers. Indeed, the surplus of lawyers might also be seen as contributing to the deprofessionalization; new admissions to the bar have increased rapidly.[68] Although the legal profession may have been undermined by recent developments, Rothman, like Engel and Hall, sees this as having potentially positive consequences for society: "Professional dominance may be replaced by a narrower, more clearly circumscribed client-expert relationship that permits the exercise of skill and judgment within a context of accountability to client and public."[69]

Eve Spangler and Peter Lehman studied lawyers in diverse bureaucratized settings in order to assess the impact of recent socioeconomic and technological changes.[70] In large law firms, they found a clearly defined stratification system: junior associates, senior associates, junior partners, and senior partners, with peer supervision and training reminiscent of craft apprenticeship models. Computerized case-retrieval data banks and paralegals were used to promote efficiency and delegate more routine tasks. Day-to-day working conditions were highly autonomous, and no complaints with management were expressed; rather, it was *client* constraints that were the focus of grievances. One lawyer said:

> In practice, you do what walks in the door. There are limits to how much you can investigate by-ways, the history of a client's problem, which may intrigue you. You must always distinguish between intellectual interests and client's needs because the clock is always ticking.[71]

Another spoke of the "banalities and peccadillos of clients in the real world."[72] Although the actual constraints (time, case load, etc.) may not derive from clients, it appears that, as with physicians, the clients are often the focus of blame.

Similarly, among lawyers working as house counsel for corporations, Spangler and Lehman found few conflicts between the lawyer and management, and a general similarity of interest. Legal work was informally organized, and lawyers worked autonomously but with managerial control of goals; this control was not viewed negatively, due to shared goals.

However, among lawyers working for government agencies, particularly those agencies serving the poor, Spangler and Lehman found a different story. Working conditions were poor, case loads were high, and dissatisfaction, burnout, and turnover were high. They found that legal services offices, under budgetary constraints, were forced to limit intake to approximately one-tenth of demand, substituting "self-help" packets for legal aid when intake was closed.[73] Paralegals and secretaries were forced to take on case-work duties in the face of overwhelming demand. Legal services lawyers also appear to retain considerable altruism in the face of such adverse working conditions, blaming management rather than clients, and turning to unionization in some cases in order to improve working conditions.

Spangler and Lehman conclude that generalizations about changes in the legal profession are difficult, given the wide range of experiences, which are dependent on setting.

Some lawyers celebrate their circumstances. This is especially characteristic of "Wall Street" lawyers who, having made their way from Ivy
League colleges to prestigious law schools to top-drawer law firms, find
their work lives not only satisfying but glamorous. . . . . Other lawyers do
not celebrate their circumstances, but they accommodate themselves to
them. Corporate staff attorneys typify this response, with low turnover
rates and generally low-key but satisfactory relationships with their
clients. Still other lawyers flee their work environment, as evidenced by
the high attrition among government lawyers. And yet others, though
still very few, choose to fight for greater control over their work lives, as
demonstrated by unionization campaigns among legal service attorneys.[74]

Not only lawyers, but also judges, are affected by recent changes in
professionalism. Diverse alternatives to conventional criminal adjudication have emerged in response to such problems of the criminal justice
system as "delay, overcrowding, high cost, offender recidivism, and
public alienation."[75] The argument for adjudication alternatives is that

our criminal courts, patterned on an adversary model for the resolution of
social conflicts, are an imperfect—and often inappropriate—societal
response to the processing of many offenders, especially those charged
with minor criminal offenses or offenses involving no substantial factual
disputes. In many lesser criminal cases the process of conventional adjudication may be too time-consuming, too expensive, somewhat irrele-
vant to, or even inconsistent with, achieving effective dispositions.[76]

Although the judicial alternatives are various, they can be grouped
into three main categories: (1) the use of nonprofessional judges (e.g.,
justices of the peace), (2) the substitution of "administrative adjudication" for judicial adjudication in areas such as traffic and housing,
thereby bypassing the conventional criminal process and decriminaliz-
ing certain offenses, and (3) the "diversion" of certain cases from the
conventional process and the substitution of a negotiated disposition for
the conventional adjudication process (e.g., "plea bargaining").[77]
    The last type of alternative has been the most widespread and the
most controversial. One reason is that diversion of cases relies heavily
on the discretion of the police, attorneys, and judges in order to determine which cases will be diverted and the nature of the disposition.
Aaronson et al. call for regulation of judicial alternatives to insure that
discretion is exercised responsibly, but the nature of such regulation is
left vague.[78]
    A fuller analysis of these and other changes in the judiciary is pro-

vided by Wolf Heydebrand.[79] He argues that in advanced capitalist countries, fiscal crises of the state (a structural gap between state revenues and expenditures) have created a crisis of the judicial system that has led to the institutionalization of technocratic administration. He defines technocracy in the following way:

> Technocracy can be defined as a system of social control based on scientific-technical knowledge and instrumental rationality in decision-making. It involves highly systematized and codified forms of knowledge ("science"), and their systematic application in terms of technology, social engineering, information processing, decision-making, and work procedures. It implies the control of all economic and political resources of the social system by scientists and engineers (the "technocrats"), and it entails the ideological and practical consolidation of social control in terms of social systems theory and comprehensive system-wide planning so as to reduce complexity and uncertainty and to maximize predictability and stability.[80]

Similarly, in their study of the federal district courts, Heydebrand and Seron found that

> courts are changing from professionally and collegially controlled, semifeudal domains of judges, to modern businesslike, administrative agencies concerned with speed, efficiency, productivity, simplification, and cost-effectiveness in the delivery of judicial services. Adjudication, under the aegis of due process and adversary procedure, is moving toward case-management, plea bargaining, and informal negotiation within an organizationally integrated system based on technical-managerial expertise and computerized information technology.[81]

The specific technocratic strategies that Heydebrand highlights in analyzing the judiciary are (1) combined centralization/decentralization of court systems, with reform movements toward centralization mitigated by sufficient decentralization to provide flexibility; (2) the adoption of technical innovations, particularly the centralized data banks of crime statistics and specific information on past offenders that provide allegedly objective data to prosecutors, judges, and probation officials to aid in disposition and sentencing decisions; (3) role integration of professional and administrative functions, with professional discretion becoming increasingly specialized, circumscribed, and "bureaucratized"; and (4) the integration of court procedures, involving the circumvention or suppression of the adversary process and the "transformation of judicial proceedings into a single, unified systemic process of decision-making and review."[82] For Heydebrand, this latter strategy not only violates norms of due process, but threatens to turn judges into what Weber conceived of as "vending machines": "Pleadings are inserted together with the fee . . . [the machine] then disgorges the judgment together with its reasons mechanically derived from the Code."[83]

However, it is not clear whether judicial discretion is being expanded or undermined by recent transformations of the judiciary. Although Heydebrand concludes that technocratic administration constrains judicial decision making, Freidson contends that recent reorganization of the judicial process has enhanced judicial discretion: "The ultimate authority of the chief judge over the court system remains supported by law, and 'by definition the [trial] judge must exercise larger amounts of discretion.'"[84] Dreyfus and Dreyfus also emphasize the ongoing salience of judicial discretion and minimize the impact of rationalization on judicial decision making.

> It is ironic that judges hearing a case will expect expert witnesses to rationalize *their* testimony, but when rendering a decision involving conflicting conceptions of what is the central issue in a case and therefore what is the appropriate guiding precedent, judges will rarely if ever attempt to explain their choice. They presumably realize that they know more than they can explain and that ultimately unrationalized intuition must guide their decision-making.[85]

In a recent review of alternatives to conventional adjudication, Aaronson et al., however, simultaneously stress both the fact that such alternatives (e.g., diversion of cases) rely heavily on the discretion of various professionals (particularly judges) and the fact that such discretion must not remain unregulated.[86]

There is then, some ambiguity concerning the extent to which recent changes in the legal profession have undermined professional discretion and autonomy; it appears that these changes have created some threats to professionalism, but the actual extent of the erosion of traditional legal and judicial professionalism remains controversial, probably because it is variable and dependent on setting.

THE EDUCATIONAL SYSTEM

The educational profession, more than medicine or law, has a long history of bureaucratization, professional/bureaucratic conflicts, and battles between centralized control and local autonomy.[87] By the late nineteenth century, a system of American education was seen as desirable, and centralization and bureaucratization were emphasized to further this goal. Business interests, which were widely influential in shaping the educational system, sought to model schools after factories, emphasizing punctuality, order, standardization, docility, and industry.[88] Demands for greater professional autonomy, among both secondary school teachers and university professors, soon began to come into conflict with bureaucratic control.[89]

The response of administrators to such conflict was to attempt to transform the early, mechanistic bureaucracies into more rational, professional bureaucracies. Earlier metaphors of factory or machine gave way to images of corporate administration by the turn of the century.[90] The new, streamlined educational bureaucracy emphasized standardization of requirements and grading procedures (particularly in colleges and universities), Taylorist attempts to objectify educational outcomes and increase system efficiency, and objective mechanisms for differentiating students into different tracks and schools.[91] By the 1920s, standardized testing was being developed and heralded as a "scientific" way of sorting pupils.

In recent years, the rationalization of the educational system has continued, fueled by new educational technology, a perceived crisis of the system, and a quest for "computer literacy" that approaches the fervor of the post-Sputnik era. Earlier metaphors of factory and corporation have given way to a desire on the part of school administrators to emulate technologically advanced industry.[92] Computerization is emphasized for both instruction and administration.

> Two potent forces are pressing schools and school systems to purchase and use computers for instructional and administrative tasks. First, federal and state politicians are demanding improved student performance, often through the use of standardized curricula and greater student testing. Second, local educators and lay boards are demanding improved student performance and tighter control over schools.[93]

Recent changes, then, are having a major effect on the teaching profession, at all levels.[94]

At lower levels, computers are affecting teaching in several ways. Computer-assisted instruction (CAI) allows for direct instruction by the computer, drill and practice routines, educational game playing, and the training of students in programming. Computer-managed instruction (CMI) is used to monitor the students' progress, in terms of work completed and skills mastered, and therefore can also be used by school administrators to monitor the effectiveness of the teacher and to specify how students should be grouped for instruction.[95] Thus, CAI can be used to substitute for teachers or to allow teaching assistants to take over many teaching tasks; CMI can be used to expand administrative control and reduce teacher autonomy.[96]

There are both benefits and dangers associated with the computerization of the classroom. The classroom computer is potentially a valuable tool for students and teachers alike, allowing for sophisticated simulations, games that provide enjoyable learning environments, and opportunities for students to work together on problems or essays through computer networking.[97] Computerized instruction can also promote "competency-based learning," or the ability of students to learn at their own pace. However, when the computer is used as a replacement for the teacher, the issues become more controversial.

> Behind the hope that computers can aid or even replace teachers is the idea that the teacher's understanding of both the subject being taught and of the profession of teaching consists in knowing facts and rules, the job of a teacher being to make the domain-specific facts and rules explicit and convey them to the student, either by drill and practice or by coaching, depending on the complexity of the subject to be taught. If that were indeed the way the mind works, the teacher could transfer his facts and rules to the computer, which could replace him as drill sergeant and coach. But since understanding doesn't consist of facts and rules, the hope that the computer will eventually replace the teacher is fundamentally misguided.[98]

Dreyfus and Dreyfus see computers as capable of assisting *training* but not *education*, since the latter depends on subtle techniques of contextualization and draws on the student's existing knowledge in order to expand that knowledge. In computerized classrooms, "the danger is in trying to teach only what can be rationalized,"[99] so that students may lose the ability to see connections for themselves and to make their own choices.

In higher education, the teaching profession and the educational system have also changed in recent years. University professors have historically been more successful than primary and secondary school teachers in preserving their autonomy and discretion in the face of administrative rationalization and control.[100] Systems of university governance have been characterized as "dual authority" or "shared author-ity," in which administrative power parallels that of the faculty.[101] How-ever, by the late 1970s this conceptualization was being challenged.

> Higher education is in the throes of a shift from informal and consen-sual judgments to authority based on formal criteria. There have been changes in societal and legislative expectations about higher education, an increase in external regulation of colleges and universities, an increased emphasis on managerial skills and the technocratic features of modern management, and a greater codification of internal decision making pro-cedures. These changes raise the question of whether existing statements of shared authority provide adequate guidelines for internal gover-nance.[102]

Changes in the contemporary system of higher education parallel broader changes in the capitalist economy. With the expansion of enroll-ments during the 1960s and 1970s, higher education became increas-ingly stratified, with community colleges centered around technical training expanded in number, providing alternatives to classical lib-eral arts education.[103] The system of higher education has been brought into closer correspondence with the capitalist economy and increas-ingly subject to similar methods of organization and control. Educa-tional administration, which has grown enormously, has eroded stu-dent and faculty traditions of self-governance, and professorial autonomy is undermined by the "use of educational technologies to distribute 'learning opportunities.'"[104] As with other professions, there appears to be an internal stratification within professorial ranks: "Highly paid, tenured faculty working at `upper-tier' institutions are thus becoming a `labor aristocracy' in relation to the growing numbers of unemployed, part-time, or untenured faculty."[105]

As in other professions, a technocratic administration of the edu-cational system is emerging.[106] Five interrelated features are character-istic of technocratic administration. First is the increased systematiza-tion, stratification, and centralization of educational institutions. Within the worldwide educational system, local autonomy is eroded by cen-tralized control by agencies, boards, and commissions. Second is the increased quantification of educational "inputs" and "outputs," with the aid of standardized tests and computerized data processing. Creden-tialism has become tantamount to commodification of the educational process and is justified as a step toward rationalizing the relationship between educational attainment and occupational allotment.

Third is the development of both educational and administrative technology. Computerized systems of administration have tended to augment centralized control.[107] Computerization also has transformed college classrooms, particularly in some disciplines. Video technology has dramatically expanded the scope of some lecturers, although it has not been as widely implemented as once predicted.[108]

Fourth is an increased emphasis on long-term planning and allegedly objective decision making based on technical imperatives. The state becomes more involved in planning commissions whose politics are veiled by the emphasis on facts, computer modeling and simulation, and alleged motivation to achieve efficient allocation of scarce resources.

Finally, a new professionalism is emerging, as the profession of teaching is transformed:

> Teachers are becoming less professionals and more 'experts,' whose autonomy is restricted but whose tasks are simplified by the increased use of machine and computer technology and standardized teaching materials. Such technology serves to effect a convergence between administrative and professional (teaching) functions, since administrators and teachers alike become adjuncts of technological rationalization. Teachers function less in order to teach and more in order to differentiate students, and they are amply aided by the appropriate technology. Moreover, technocracy serves to blur the lines of power in the educational hierarchy, since on one level teachers are expert administrators of advanced technology . . . while on another level they are mere functionaries implementing a technology which is centrally devised and controlled.[109]

One illustration of this convergence of administrative and professional functions is the 1980 Supreme Court decision whereby professors at Yeshiva University were declared to be managerial employees of the university and therefore ineligible to engage in collective bargaining.[110]

The Yeshiva case aptly reflects the ambiguity inherent in the role of university professors today. Faculty and administration are being integrated within the technocratic system. However, the process is a contradictory one, with faculty and administrative interests not yet identical. Certainly faculty continue to exercise considerable autonomy, discretion, and power within universities, but the nature of their work has changed considerably in recent years, notably by the subordination of their teaching function to a managerial, differentiating function. To proclaim administrators and professors as a unified, expert coalition, united in their goal of "managing" students and nonprofessional staff, will not resolve the ongoing conflicts of interest within contemporary colleges and universities, and may in fact exacerbate them. It is these conflicts between administrators, faculty, students, and staff that will determine the future shape of the university and the educational system.

HIGH-TECH PROFESSIONS

Finally, we will consider the professional and paraprofessional jobs associated with advanced computer technology: computer engineers, systems analysts, computer programmers. Traditionally, engineers have had less autonomy than the more elite professions because engineers have been more closely allied with corporate interests.[111] With the increasingly complex and scientific nature of production technology, have engineers gained power and importance, or has capitalist control of their activities become more stringent? What are the working conditions and social organization of work like for the high-tech descendants of traditional engineers?

Haug contends that much traditional engineering expertise has been automated, or replaced by sophisticated computer programs such as computer-aided design (CAD), leading to deskilling and the paraprofessionalization of engineering.[112] Social scientists who have empirically studied the high-tech professions, however, reveal a rather more complex picture.

Harold Salzman, for instance, points out that much of the negative evidence regarding the impact of computerization on engineering is taken from vendors' advertising materials or managerial predictions of the *probable* effects of new technology; while indicative of a certain potential, and of managerial intent, such predictions may not accurately reflect the actual effects of the technology.[113] His empirical findings, based on extensive interviews with CAD users (in this case, designers of electronic computer circuits) and managers in eight high-tech companies, as well as on participant observation at the workplace and at trade shows, reveal that such intentions are not yet realized, largely because the technology does not perform in accordance with the advertising hyperbole.

In general, Salzman found that upskilling rather than deskilling accompanies CAD, and he also found no evidence of any erosion of professional autonomy. Even if managers had considerable technical and theoretical training, he found that they tended to rely heavily on the skill and experience of their designers and to grant them considerable autonomy. To overly control skilled workers was seen as counterproductive in terms of efficiency of production. Hodson also found that the high-tech engineers he studied enjoyed substantial autonomy from managerial control.[114]

Shaiken also studied CAD and found both positive and negative potential in it.[115] He sees enormous advantages in terms of expanding design alternatives, enhancing the capacity to make revisions, and widening the possibilities of professional collaboration. Shaiken analyzes the organizational impact of CAD and concludes that it promotes both vertical and horizontal integration of the production process. Vertical operations, from design to actual production, are easier to link, as are the various engineers themselves.

> On a large project, different teams of engineers feed their design work into the data base, which makes it instantly available to the other groups. Whether there are ten designers or a thousand, they all have the ability to view the same information at the same time. This makes it possible to work on different parts of an assembly in parallel without worrying about whether it will all "fit" when complete. Moreover, the design engineers are constantly able to interact with the manufacturing engineers, who determine how the part is made.[116]

When telecommunications is combined with CAD, worldwide networks of engineers can work on the same project.

However, despite this positive potential, Shaiken sees the actual implementation of CAD as more negative. Most problematic is the further separation of conception from execution, as the design process is increasingly separated from the shop floor and computer models are abstracted from reality. Shaiken's concern over this trend is both practical and political: practical due to the increased probability of error, and political because of the social implications of divorcing intellectual and manual work.[117]

Shaiken sees both deskilling and upskilling as potential effects of CAD. Like Salzman, he sees the software as primarily involved in lower-level decisions and tasks. Unlike Salzman, however, he sees the possibility, given the specialization and stratification of the profession, that upper-level and lower-level tasks may be assigned to different people. CAD may therefore have different results on different levels: "Although for senior designers, the application of CAD may open new horizons, for many others, intellectual speed-ups and mental hazards result in new forms of stress. . . . the role of the junior designer may become an entirely clerical one, devoid of any decision-making content."[118]

Philip Kraft has studied computer programmers and draws similar conclusions about stratification within the profession and contradictions between the potential and the reality of technological innovation.[119] Like Salzman, Kraft sees managerial intentions of enhancing their control of high-tech workplaces through deskilling and specialization. Unlike Salzman, however, Kraft concludes that such intentions serve to thwart the positive potential of advanced technology.

Like Shaiken, Kraft contends that polarization is occurring within high-tech professions and that advanced technology has widened the breach between top and bottom. Coders and low-level programmers perform clerical or paraprofessional work, whereas senior programmers, computer engineers, and systems analysts retain more autonomy and discretion in their work. Managerial control devices, such as structured programming, impact most heavily on low-level workers, routinizing their work and making it easier to monitor and control.

> Structured programming, in short, has become the software manager's answer to the assembly line, minus the conveyor belt but with all the other essential features of a mass-production workplace: a standardized product made in a standardized way by people who do the same limited tasks over and over without knowing how they fit into a larger undertaking.[120]

Kraft also analyzes the "chief programmer team," which he found to be an increasingly prevalent way of organizing high-tech professions. Such teams consist of an upper-level "chief programmer," with various auxiliary junior programmers, coders, and so on. This form of team organization is therefore extremely hierarchical and status-bound. Work can be delegated and supervised by the chief programmer with the aid of structured programming.

According to Kraft, the nature of professional work has been fundamentally transformed in the high-tech professions. The polarization has eroded traditional apprenticeship models of training, with formal credentials being the norm. Formerly fluid job categories have been fragmented, with a proliferation of slightly different job titles providing horizontal mobility as an alternative to the upward mobility prospects that have been lost. For most high-tech workers, professionalism has become little more than a managerial ideology.

There is little agreement, in these times of rapidly changing work and workplaces, what, exactly, professionalism is. There is no doubt, however, what programmer managers mean by professionalism. It is a code of behavior to be adhered to by programmers but drawn up by their employers. The traditional features of professional organization and ideology—peer control of admission, training, prices, and sphere of power— are omitted and replaced with an inventory of acceptable and unacceptable behavior. Properly internalized by employees, such a code relieves managers of the necessity to directly supervise . . . software workers. An internalized "professional" ideology, in other words, replaces the conveyor belt and the foreman.[121]

Kraft also found that the polarization within high-tech professions was correlated with class background and with sex. The role of women in software work, in particular, he sees as a "social litmus test of the claims made for the impending information society and the knowledge workers who will populate it."[122] High-tech firms fail this test, as he found that women were segregated into low-level, low-paying jobs, paid less even when in upper-level jobs, and generally marginalized within high-tech firms.

Others who have investigated the role of women in computer engineering professions have also documented the fact that high-tech is not gender neutral.[123] Myra Strober and Carolyn Arnold found that women represented less than a third of programmers and systems analysts in the United States in 1980 and that women and racial minorities were overrepresented in nonprofessional jobs, but considerably underrepresented in professional jobs, particularly upper-level professional jobs. Like Kraft, Strober and Arnold found that women were paid less than men at all levels.[124] They conclude that "the computer field was sired by the fields of mathematics and engineering, and the newly born prestigious and technical jobs quickly took on the gender designation of the parent fields."[125]

Sally Hacker has amply documented the way in which engineering as a profession is gendered.[126] She argues that "engineering contains the smallest proportion of females of all major professions and projects a heavily masculine image that is hostile to women."[127] The engineers she studied told her that engineering was a male activity: hard, clean, predictable, abstract, technical, mathematical, controlling. Disciplines such as the humanities and social sciences, on the other hand, were seen as "womanly": "soft, inaccurate, lacking in rigor, unpredictable, amorphous."[128]

Hacker found that in engineering classes, the male professors often used jokes to put down women, as well as bodily functions, racial minorities, and the technically incompetent. Engineering magazines often included pictures of nude or partially nude women, and one included both a centerfold ("E-girl of the month") and a dirty joke page.[129] Engineering was perceived as a male fraternity, with mathematics courses (in particular) serving a gatekeeping function. One statistics professor, for instance, when asked why students could not have more time for an exam, said, "If we gave people more time, anyone could do it. The secretaries could even pass it."[130]

The male engineers she interviewed also reported limited experiences with intimate, close personal relationships, difficulties in expression emotion, and a sense that sensual pleasures were "unmanly." Both as children and as adults, the men had experienced estrangement from girls and women, feeling shy, insecure, frightened and mystified by the female sex. Largely because of the childbearing ability of women, these men associated women with nature. These male engineers also reported painful childhood experiences of insecurity over lack of athletic ability and self-perceptions as "sissies" and as sickly and lacking in courage. What emerges is a pattern of compensation through academics, compensation for lack of other masculine traits; as one man put it, he tried to compensate by being "head of the class."[131]

Engineers were considered to be the most appropriate men for managerial positions "because they can treat people like elements in a system."[132] Engineering was seen as closely allied with ideologies of control and dominance: control of nonexpert workers, women, racial minorities, the body, and nature in general.[133] The abstract rationality of engineering rests on a mind/body dualism and a strong sense of hierarchy: The reward for denial of the body, emotionality, and intimate relations is social privilege. It becomes apparent, however, that the male engineers themselves are also stringently controlled and limited by this discourse.[134]

An ethnographic account of the nongendered dimensions of engineering culture, and the way in which organizational culture can be used as a type of control, is provided by Gideon Kunda.[135] In the high-tech engineering firm he studied, he found three main categories of workers: exempt workers (managers and engineers), nonexempt workers (secretaries and clerical workers), and temporary workers (who receive no benefits and are not really considered to be employees of the firm at all). Particularly among the exempt workers (analogous to my expert-sector workers), Kunda found that a distinct organizational culture was used as a type of control.

The official guidebook of the firm, for instance, describes this cultural ideology in the following way:

> High Technologies is a people-oriented company. The employees receive courteous, fair and equitable treatment. . . . Management expects hard work and a high level of achievement. . . . A great deal of trust is placed in employees to give their best efforts to a job. . . . The matrix organization is goal-oriented and depends on trust, communications, and team work. As a result, most employees function as independent consultants on every level, interacting across many areas necessary to accomplish the task.
>
> Honest, hard work, moral and ethical conduct, a high level of professionalism, and team work, are qualities that are an integral part of employment at High Technologies. These qualities are considered part of the Tech culture. Employees conduct themselves in an informal manner and are on a first-name basis with everyone at all levels. . . . The opportunity for self-direction and self-determination is always present.[136]

Kunda found that this ideology served as a type of control that was enforced by a broad group of control agents, which included virtually all employees, but particularly the expert-sector workers.[137] This cultural control thus becomes a deeper and more insidious type of control, one that is internalized: "Hard work and deference are no longer enough; now the 'soulful' corporation demands the worker's soul, or at least the worker's identity."[138]

One of the most salient characteristics of the firm Kunda studied is its polarization. Only the exempt workers are considered to be full citizens of the firm; the nonexempt workers and temporary workers are ostensibly included in the rhetoric of the official ideology, but these workers view their inclusion with skepticism. In fact, Kunda argues that the nonexempt workers and temporary workers are subject to utilitarian and coercive control rather than the normative control of the exempt workers. In effect, there is a pronounced polarization into "central" and "marginal" workers, with very different types of control for each group.

> One consequence of the managed culture is a polarization of central and marginal members: the former must accept or limit an extensively defined organizational self; the latter must live with or seek to enlarge a minimal one.[139]

Kunda concludes, in fact, that the marginal workers become, in effect, "non-persons," although the cultural ideology of community attempts to obscure this devaluation.[140]

CONCLUSION

Recent empirical work on changes in professionalism reveals both support for and refutation of the theses of deprofessionalization and proletarianization. It appears that these conceptualizations, while somewhat exaggerated, are nonetheless of heuristic value in understanding recent changes in certain professions.

Advanced technology and other rationalization measures appear to have differential effects on professional work, depending on the relative status of the profession and of the professional within a certain profession. Several analysts have pointed to the increased stratification and underlying polarization of the professions.[141] Given this polarizing tendency, it is not surprising that technological and organizational changes are having divergent effects. The National Research Council's conclusion about advanced medical technology has broader relevance:

> The possible employment effects of such systems are, however, hardly spelled out. As is the case for other expert systems, there are possibilities for raising the professional standing and knowledge of one's own group as well as possibilities of diffusion of knowledge to groups with less formal education. The latter effect can lead to replacement or curtailment of growth of the "expert" group through employment of groups with less formal training. Systems planned as labor-saving devices might turn out to demand increased input of labor of another kind, either more or less professional. . . . Whether groups of workers perceive the introduction of new technology as a threat to be averted or as a challenge to further their own interests will depend partly on their knowledge base both about their own field and about the technology in question.[142]

Elite professionals within elite professions (most physicians, many lawyers, most university professors, judges, systems analysts, computer engineers) have undergone a transformation of their working conditions, but not deprofessionalization. Although a certain degree of autonomy may be lost as these professionals become more integrated into complex administrative systems, considerable discretion over day-to-day professional tasks is retained. Ideologically and economically they may be more subject to capitalist and bureaucratic imperatives, but this subordination does not appear to cause conflict, due to general similarity of goals. Advanced technology generally appears to enhance rather than undermine professional work for these groups, eliminating their more routine tasks, expanding their data base, and improving their decision making.

Less elite professionals (health care technicians and nurses, teachers, computer programmers), however, appear to be more vulnerable to professional rationalization.[143] They may be displaced or deprofessionalized by technological innovation. Their work may be deskilled, their case loads may be increased, their contact with clients may be routinized, their work may be more stringently controlled and monitored. For such workers, professionalism may become little more than a legitimating ideology, with ideological differences being their primary source of differentiation from nonprofessional workers. Moreover, given the sex and race segregation which has been found to be characteristic of high-tech and other professions,[144] this polarization of the professions has clear gender and racial implications.

Chapter 6

EMERGENT TECHNOCRACY

We have seen that diverse work organizations are being restructured around advanced technological systems. Production workplaces, bureaucracies, and professional organizations have been transformed by managers and technical experts in ways that parallel one another, leading to the emergence of a new type of organizational control structure: technocracy.

Technocratic organization is most apparent in workplaces centered around computerized technology: superautomated factories, high-tech firms, telecommunications companies, insurance companies, medical systems. The correlation between technological innovation and organizational innovation appears to derive from two facts: (1) High-tech research and development firms and other technologically advanced organizations usually operate in highly competitive markets emphasizing product innovation and therefore seek alternatives to bureaucracy.[1] (2) The installation of advanced technological systems, even in more "mature" firms and industries, opens up opportunities for social and political restructuring; some, indeed, have argued, that such advanced technological systems cannot be fully realized without such organizational restructuring.[2]

Advanced technological systems are *flexible* but not *neutral*. They can be (and have been) designed and implemented in different ways, but once these choices have been made, the social and political effects on workplace organization tend to be lasting. In practice, the organizational flexibility associated with advanced technology has not yet been widely realized, and technocratic organization, while not monolithic, is assuming specifiable features. In this chapter we will delineate the specific aspects of emergent technocratic organization and control, as well as certain social and political implications.

CHARACTERISTICS OF CONTEMPORARY TECHNOCRACY

Polarization

Polarization into expert and nonexpert sectors is a fundamental aspect of technocracy. As organizations are systematically reorganized around computerized technology, which substitutes technological complexity for the organizational complexity characteristic of bureaucracy, the organizational structure is transformed. The hierarchical division of labor is simplified and broken down, with a polarization into expert and nonexpert sectors, the elimination of many middle-level positions, and a flattening of the occupational hierarchy. In diverse workplaces, an expert sector, composed of managers, technical experts, and professionals, is clearly differentiated from a nonexpert sector, whether of clerical or production workers (or both).

Moreover, this polarization is paralleling and reinforcing race and sex segregation, with white males predominating in the expert sector, and women and racial minorities disproportionately found in the non-expert sector.[3] Although in recent years some women have gained representation in expert ranks, particularly in certain fields, the general pattern of sex segregation persists, with clerical work, for instance, becoming even more female dominated.[4] Although computerization has been accompanied by feminization, the usual trend has been for women to gain representation in nonexpert-sector operator jobs with very limited mobility prospects.[5]

The use of computerized technology fundamentally alters the separation into conception and execution characteristic of mechanical technology, as the technology takes over much of the actual execution.[6] In superautomated factories, for instance, the new dichotomy is between conception and *monitoring* of the computerized technology. Effective monitoring requires certain conceptual, diagnostic skill and some degree of comprehensive understanding of the technical system, which is prone to periodic technical failures. Management therefore becomes more (rather than less) reliant on nonexpert-sector workers.

If workers are given opportunities to gain comprehensive knowledge and exercise conceptual skill, the dichotomy between conception and execution can be mitigated. However, existing evidence indicates that management often resists allowing workers to gain such knowledge, due to managerial control prerogatives, preferring to rely on technical experts to design and maintain the technological system.[7]

As certain bureaucracies are transformed into technocracies, different types of polarization are apparent. In workplaces (or divisions) where the type of work is sufficiently standardized to be able to be largely assumed by the computer system, the polarization is most extreme: The nonexpert sector of data entry clerks is widely separate from the expert sector of managers and technical experts.[8] However, even when the nature of the work resists such complete automation, as in insurance firms with their need for personal interaction with clients, the focus on task reintegration around skilled clerical workers, which some firms have adopted, nonetheless results in a bifurcated structure: expert workers and skilled (or semiskilled) clerical workers. Similarly, in the newspaper industry, as professional journalists and editors assume word-processing functions once assigned to compositors or clerical workers, the number of nonexpert-sector workers is reduced but not eliminated; workers in the pressroom and the mailroom comprise a nonexpert sector that is clearly separate from the expert sector.

This trend toward polarization has several corollaries. One is that internal labor markets and seniority-based promotions are being deemphasized and that external credentialing and stringent credential barriers between the levels of the flattened hierarchy are becoming more prevalent.[9] A proliferation of job titles within the nonexpert sector may veil the underlying polarization and give the opportunity of horizontal mobility as an alternative to upward mobility. "Career ladders" thus become reduced to minute gradations among nonexpert jobs.[10] Recruitment is typically from without the firm, and upward mobility within technocratic organizations is either nonexistent or achieved via external credentialing, an option that is not widely available to nonexpert-sector workers.

Another corollary of polarization is that the organization and nature of the labor process vary dramatically according to one's level in the flattened, polarized hierarchy. Expert and nonexpert jobs differ not only in their level of skill and remuneration, but also in terms of work organization and working conditions.

At expert levels, much of the rigidity characteristic of bureaucratic rules and task specifications tends to be replaced by more flexible and collegial types of work organization. The distinction between professional and managerial jobs is blurred, as professional and managerial tasks are often combined, either within task forces or within the same person; *manager* has come to mean someone with high status, rather than someone who manages or supervises others.[11] Unlike professionals working in bureaucratic enclaves, professionals in technocratic organizations are more fundamentally integrated into the organization. Tasks tend to be organized around ad hoc projects, with the formal structure of communication and chains of command routinely bypassed, leading to more decentralized authority among the expert sector, an emphasis on horizontal communication, and a sense of "adhocracy."[12]

Power, privilege, and influence are also allocated differentially according to sector within the bifurcated technocratic workplace. Expert-sector workers are likely to have considerable influence on managerial decision making, whereas nonexpert-sector workers tend to be restricted to more superficial types of participation.[13] Expert-sector workers also benefit greatly from computerized production and information systems, which enhance their work by providing them with expanded access to relevant data, enlarged collegial networks (both intra- and interfirm), and technological support for conceptualization and decision making. Indeed, telecommunications systems now link experts around the world, making it possible for experts such as engineers to work simultaneously on the same project.[14]

At the nonexpert level, the tendency has been for tasks to become more routinized and/or stringently monitored through the technology.[15] However, the nature of nonexpert-sector working conditions depends upon social and political choices, constrained but not determined by technology. Advanced technology can also be implemented so as to enhance nonexpert work: to "increase employees' feedback, learning, and self management rather than to deskill and routinize their jobs or remotely supervise them."[16] In some organizations, nonexpert workers have been organized into multiskilled teams, with opportunities to learn and acquire comprehensive understanding of organizational and technological functioning.[17]

However, superficial types of worker participation may mask managerial control or expert privilege.[18] For instance, when teams involve workers of heterogeneous status, external rank and knowledge disparities have been found to adversely affect team functioning.[19] Non-expert workers have sometimes felt "dumb" or manipulated in such contexts. In worker teams where nonexpert-sector workers have been given more opportunities for genuine participation, technical experts have sometimes protested that the decisions made were not the "best" ones.[20] Even though workers in technocratic organizations often work more cooperatively, this does not necessarily imply the transcendence of the expert/nonexpert polarity. Given the strong tendency toward polarization, undermining the polarity between expert and nonexpert workers has proven to be a difficult task.

Studies of recent changes in professionalism indicate that status differences and a certain polarizing tendency have also become more apparent within the ranks of professionals in recent years. Freidson speaks of an emerging administrative elite, a knowledge elite, and rank-and-file professionals, a typology in which is embedded an underlying polarization into elite and rank and file professionals.[21] Similarly, Spangler and Lehman analyzed stratification within the legal profession, finding a wide disparity between the world of Wall Street and corporate attorneys, on the one hand, and lawyers who work for government agencies, particularly those serving the poor, on the other hand.[22] High-tech professions are among the most obviously polarized, with coders and low-level programmers performing increasingly standardized tasks and senior programmers and systems analysts retaining considerable autonomy and skill in their work and substantial power within the organization.[23]

Medical systems are a good example of technocratic professional organizations. Sophisticated and capital-intensive medical technology, in conjunction with socioeconomic changes, have led to dramatic changes in health care. Paraprofessionals and technicians are becoming more numerous and more integral to health care practice. Increasingly, health care is technologically mediated, and the evidence indicates that advanced medical technology affects physicians and paraprofessionals very differently. Physicians in effect become expert administrators of the technological apparatus, using it as a tool to enhance their diagnoses and treatment decisions.[24] Medical technicians and paraprofessionals, on the other hand, find their jobs more limited in discretion, more routinized, and less client-oriented as the technology has become more advanced and the work load has become heavier.[25] Nurses, however, have generally been able to resist such job degradation, even though the technological potential for more extensive automated patient monitoring exists, due to social preferences (of doctors, nurses, and patients) for more personalized nursing care. Technological change, although influential in modern health care systems, is not deterministic, but rather operates within a clear social and political context.

The technocratic organization of professional care therefore involves increased technological dependence, specialization, and stratification. An underlying polarization into elite and less elite (or paraprofessional) professionals is also apparent in many contexts. In organizations that rely on paraprofessionals (e.g., nurses), the overall polarization tendency is mitigated, as an expert sector, a "paraexpert" sector, and a nonexpert sector are apparent. However, in other contexts, paraprofessional work may be sufficiently undermined by the sociotechnical restructuring to effectively render it nonexpert.

Centralization/Decentralization

Another major way in which technocratic control differs from previous forms is that centralization is combined with decentralization in varying configurations.[26] Traditionally, organizational control has been kept highly centralized, but while centralization of control is the norm in technocratic organizations, it is not inevitable; centralization is combined with varying degrees and types of decentralization.

In both blue-collar production workplaces and corporations, microcomputers have opened up new opportunities for decentralization. Computerized numerical control (CNC), for instance, can be used to facilitate worker programming and editing, and microcomputers can serve to provide information and communication links to relatively autonomous satellite stations or work teams.[27] However, the same technology can also be used to enhance centralized managerial control, surveillance, and monitoring of worker performance, and the evidence indicates that the more centralized model is currently the norm.[28]

It appears that firms that adhere to a more centralized model of computer automation also tend to rely on a more specialized division of labor and more stringent technical control. However, such centralized systems that stress technical control can backfire in terms of productivity, not only because of higher turnover rates, but also due to worker strategies that foil the computerized monitoring process at the expense of customer service or quality production.[29]

The technocratic system often assumes the form of *visible* decentralization (e.g., a computer terminal in every office or throughout the shop floor) but with an underlying centralization (or quasi-centralization) of control that is programmed into the design of the computer system. The technology can be used both to promote and to obscure the polarization of power. In practice, computerized control tends to be kept quite centralized, although at expert (or paraexpert) levels access to power and information is more widely distributed.

That computerization has tended to augment centralized control and systematization on a worldwide scale is clear; what is less clear is the extent to which there is the opportunity for a genuine decentralization of power. Obviously, the engineering and programming of computerized systems is critical in terms of determining such political variables as the balance of centralization/decentralization or the stringency of technical control at the nonexpert level. In order to exercise significant power within the technocracy, a person must have both expertise and access to a computerized technology that has been preprogrammed to allocate such access differentially to expert and nonexpert sectors. Given the flexible programming options, the social relations of production become highly salient in computerized workplaces and interact with the technical relations of production in an ongoing manner. Existing organizational power relationships shape technological design and currently appear likely to thwart technological potentialities that might undermine technical control.

Skill Restructuring

Although deskilling is an overly simple description, "skill restructuring," "skill disruption," and new types of alienation, stress, and occupational hazards are apparent in technocratic workplaces.[30] Although both deskilling and reskilling occur with the reorganization around advanced technological systems, the balance between the two trends is determined by both the design of the technology and the way in which it is implemented. Such factors as the degree of task diversity, the nature of group interaction, and the amount of worker discretion are more influenced by social and political factors. For instance, a given *task* may be deskilled by a certain technology, but depending on how many tasks are included in a given job, the *job* may be upskilled.[31] Social and political factors interact with technological development, resulting in complex outcomes.

As we have seen, more elite professionals may experience a certain loss of autonomy as they become integrated into complex technocratic systems, but the nature of their work generally appears to be enhanced rather than undermined by recent sociotechnical change. Technological support systems expand both the amount of data available for professional decision making (e.g., medical diagnosis programs, on-line court case histories) and contact with collegial networks. In some professions, computerization of the more routine aspects of a job may result in upskilling.[32] However, for less privileged workers the effects of sociotechnical restructuring on skill levels have been more diverse and ambiguous. For instance, numerical control of machining using microcomputer terminals can be implemented so as to deskill/monitor or enhance the work of the machinist. The central question concerns who programs the technology. Typically, professional programmers do the programming. However, the technology has been implemented in diverse ways: "Decisions about how to program the machine tool are choices about how to organize the workplace. They influence how much skill and control the machinist retains on the shop floor and where in the management hierarchy key production decisions are made."[33]

Whether a job is deskilled or upskilled also depends upon the skill level of one's *previous* job. For instance, in the insurance industry, skilled clerical workers now use computerized underwriting programs to perform underwriting tasks; whether the job is perceived as deskilling or upskilling depends on whether one is looking at it through the eyes of an underwriter or the eyes of a low-level clerical worker. Similarly, newspaper photocomposition appears deskilling from the point of view of traditional craft compositors, but upskilling from the point of view of a clerical worker entering the occupation for the first time. As Cockburn discovered, such perceptions of skill levels are also influenced by gender; the male newspaper compositors considered photocomposition deskilled because "girls could do it."[34] As technological innovation turns formerly blue-collar workplaces into more white-collar settings, both class and gender confusion accompany the skill restructuring.

Another aspect of the restructuring of skill is the fact that work that involves computerized systems, whether computerized production systems or advanced information systems, involves new types of analytical skill.[35] New types of abstraction and diagnostic skill are needed, particularly when systems malfunction. The increased emphasis on abstraction does not necessarily imply that computer-mediated jobs are challenging or rewarding, however; tasks can be routinized and boring while nonetheless requiring focused, abstract attention. Sociotechnical theorists contend that workers who work with advanced computer systems must be given opportunities to acquire comprehensive knowledge of the system and to forge cooperative working relations with other workers if they are to be able to adequately contend with the new tensions resulting from "the counterpoint of watchfulness and boredom."[36]

Finally, skill level is not identical to job quality. Deskilling does not necessarily imply work degradation, and rising skill levels are not sufficient to insure improved working conditions. A job may be deskilled, but social organization (e.g., worker teams and job rotation) and/or material rewards may offset such deskilling. Conversely, a job may require relatively high levels of skill, but the work may be stringently supervised and poorly paid.[37]

Skill restructuring and disruption, not only as a result of technological change but also due to sociopolitical choices, are likely to continue. As certain jobs are fundamentally altered, and others eliminated, both workplace organization and labor market opportunities will continue to change. Training and retraining programs will therefore be of increasing importance in the years to come.

Expertise as Authority

In technocratic organizations, demonstrated expertise and credential certification tend to become more important than rank position as the basic source of legitimate authority. In contrast to the more personalized managerial authority characteristic of bureaucracy, technocratic authority rests on allegedly neutral decision making derived from expertise and systems maintenance.

> Since Weber the discussion about power has gradually been replaced by theories of decision-making systems. The authoritarian leader has been succeeded by the decision-maker. Power has been diluted of its individual-emotional content and pumped into channels of information and other control systems.[38]

Although managers and technical experts continue to make political decisions, they increasingly do so behind a veil of technocratic ideology that purports to reduce politics to technical decision making. In actuality, power interacts with technical concerns in complex ways.

In organizations where rank authority is not yet based on technical expertise, and where managers may have less technical knowledge than their subordinates, conflict has been a frequent occurrence. Such managers have found that their rank authority provides insufficient legitimacy, leading to worker perceptions of managerial incompetence and ensuing problems of authority.[39]

In technocratic organizations, certain technical and professional functions tend to converge with managerial ones, and there is a trend toward assigning these tasks to the same person. If this trend toward convergence continues, the conflict between technical experts and managers may subside. As technocratic organization becomes more fully consolidated, managers will increasingly need technical expertise, external credentials, and managerial expertise in order to maintain legitimacy. Technical expertise is becoming an important precondition for traditional managerial functions: supervision, planning, marketing, and so forth. Moreover, with increased client loads and a more rationalized work situation, the work of many professionals requires administrative skill.

Given the increased emphasis on expertise as the basis of authority, new types of politicking centered around "conspicuous expertise" become apparent in technocratic organizations. Formerly, successful bureaucratic careers depended on who you knew more than what you knew. Technocratic careers may depend not only on who and what you know, but also upon how well you can *appear* knowledgeable: "Knowledge has become a critical resource in the politics of class struggle, both inside and outside the workplace."[40]

As bureaucratic rigidity gives way to more flexible task-force and team organization, the micropolitics of small-group interaction becomes salient. Such factors as technical ability and knowledge, having the "right" credentials, charisma, gender, race, and appearance have been found to influence such micropolitics.[41] It appears that in some instances democracy conflicts with technocracy in small-group interaction, as when technical experts object to group decisions on the grounds that the "best" technical solution was not chosen.[42]

Studies of public sector technocracies reveal that not only micropolitics but also macropolitics affects managerial decision making, despite increasing legitimation of decisions on technical grounds.[43] "Technopolitics" operates behind the scenes, incorporated into the assumptions that govern scientific studies and computer modeling, the ways in which findings are interpreted, and the ways in which results are reported. Technical experts and technical expertise are increasingly influential in public-sector decision making but are often used to legitimate vested political interests.

Technocratic Ideology

Technocratic control ideology rests on the increased salience of knowledge and technical expertise, with the assumption that technical imperatives have displaced traditional politics in organizational decision making and management. The idea that there is one best way to make any organizational decision has been powerfully revived by recent developments in science and technology: "Any problems are defined in technical terms and assumed to be soluble with the aid of scientific knowledge and advanced technology."[44] The belief that the correct technical solutions can only be found by the experts becomes a powerful legitimation of expert power, both within technocratic workplaces and in the overall technocratic system.

As national and international economic systems have become larger and more complex, system maintenance becomes another emphasis of technocratic managers. As we have seen, managers, technical experts, professionals, and workers are constrained by the necessity of accommodating to overall system exigencies, and yet political influences and choices are also apparent, particularly in the expert sector. The emphasis on system maintenance can eclipse other options, however, even among some expert sector workers; in medical systems, interns and residents sometimes focus on "getting rid of patients" in order to reduce their work loads and enhance system functioning, and professors "manage" their increasingly large student loads in accordance with the needs of the system.[45] Professionals working within technocratic organizations are often ideologically subordinate to system needs, although they have more autonomy than nonexpert-sector workers.

The ideology of technocratic management therefore veils political motivation and the range of choices as to the nature of the system, through recourse to the ideology of system preservation, which has more appeal in an era of clear socioeconomic crisis tendencies. Indeed, the systemic nature of technocracy is a distinctive feature.

If bureaucratic conservatism, according to Karl Mannheim, tends to "turn problems of politics into problems of administration," and if "professionals tend to turn every problem of decision making into a question of expertise" . . . then the technocratic strategy can be said to turn problems of politics, expertise, and administration into problems of cybernetic systems control, the ultimate form of the "administration of things."[46]

Moreover, intensified ideologies of technological progress and technological determinism serve as powerful sources of legitimation, also obscuring political choices. As we have seen, workers in diverse workplaces are powerfully influenced by beliefs such as "You can't stop technological progress" or that "working with computers" is inherently important and positive.[47] The fact that both expert and nonexpert workers often utilize similar hardware, and are both working with the same technological system in many workplaces, contributes to the belief among nonexpert workers that working with computers is inherently high status and important work. Even when workers deplore certain changes in their jobs, the mystique of computers can serve to offset dissatisfaction.

In some contexts, computerized technology is seen as being mysterious—almost magic. The shift to photocomposition, for instance, inspired technological awe in the compositors. Whereas mechanical technology, such as the hot-metal typesetting process, was visible and readily comprehensible, computerized production is more hidden from view. That such enormous power and productivity can operate in a small box, without moving parts, seemed wondrous—making challenges to the organizational changes that accompanied computerization more likely to be half-hearted or nonexistent.[48]

Conversely, when the restructuring around computerized technology is viewed negatively, for instance, because of stringent worker monitoring or pacing via the technology, employees may blame the technology rather than management for this control. Zuboff found that a group of bank employees protested the intense and hard-driving nature of their jobs on computer terminals but felt that their manager was "friendly, relaxed, and fair-minded."[49] There is also evidence that management may try to create a strong corporate culture and sense of community in order to promote worker allegiance and new types of ideological control.[50] Workplace reorganization is often described to workers as determined by technological imperatives—and indeed is probably often seen by management in these terms. As we have seen, the same technology can be designed and implemented in widely divergent ways, but this social flexibility is not usually acknowledged or realized.[51]

Technocratic ideology therefore exaggerates actual aspects of technocracy: the fact that technical expertise and advanced technology have become increasingly salient and important in work organizations. As with previous ideologies of organizational control, technocratic ideology alleges objectivity and meritocracy: that technical expertise and technology are independent of political or economic interests. To challenge technocratic reorganization is therefore construed as irrational or as incompetence.[52] In actuality, technocracy is far from apolitical; technology and scientific knowledge have certainly been shaped by political interests, even as they also influence the shape of politics.[53]

TECHNOCRACY AND SOCIAL STRATIFICATION

As we have seen, technocratic restructuring is having stratification effects within workplaces, most notably promoting polarization, the erosion of internal labor markets, increased reliance on technical expertise and credentialism, and sex and race segmentation. In some organizations, however, the emphasis on paraprofessional labor mitigates the overall polarization, and in many workplaces the proliferation of job titles veils it.

Technocracy also is correlated with broader stratification effects at the level of national and international labor markets. There is evidence, for instance, that the U.S. class structure has become more polarized in recent decades. From an analysis of such factors as income data, inflation effects, and home ownership patterns, Blumberg concluded that during the 1970s the U.S. class structure was characterized by divergence and greater rigidity, casting great doubt on postindustrialist optimism.[54] In a more recent analysis, Reich presents evidence that U.S. incomes diverged between 1977 and 1990, with the wealthiest one-fifth of the population having over half of the nation's income by 1990.[55] Reich concludes that the new-class stratum, which he terms "symbolic analysts" (and which is analogous to the expert sector), is diverging so rapidly from the rest of the world that this represents a virtual secession.

Another study of labor market segmentation found that the post-World War II period witnessed a divergence between both "core" (large, monopolistic) firms and "peripheral" (smaller, competitive) firms and between "primary" and "secondary" jobs.[56] Moreover, secondary labor processes have been found to be increasingly salient within core firms, in contrast to earlier formulations of dual labor markets, which assumed that core firms generally relied upon more homogeneous and advantageous labor practices: relatively high pay, job security, mobility prospects. In recent years large core firms have increasingly organized their nonexpert sector along the lines of the secondary labor market (more characteristic of small peripheral firms): low pay, little job security (some emphasis on temporary workers or subcontracting), reduced or nonexistent mobility prospects.[57] A polarization within the primary labor sector into "independent primary" jobs and "subordinate primary" jobs, which is analogous to the distinction between expert jobs and paraexpert or semiskilled jobs, has also been documented.

Some have argued, however, that the entire dual-labor-market perspective has been called into question by recent changes in organizational structure and labor markets.[58] Not only have internal job ladders, assumed to be generally characteristic of core firms, been eroded dramatically, but the very distinction between core firms and peripheral firms has been rendered problematic. Core firms assumed their privileged position within the economy (and hence were able to offer privileges to their workers) because they were situated within oligopolistic or monopolistic markets (e.g., steel, automobile, utilities, AT&T). As Noyelle points out, this socioeconomic context has changed.

> Under deregulation and internationalization, the return to price competition in many sectors of the economy has undermined considerably the capacity of core firms to continue to operate in a sheltered oligopolistic environment. As a result, a new economic fragmentation is now emerging between sunrise and sunset industries, substituting for that between core and periphery industries. *Sunrise industries* tend to develop in highly competitive environments. They include . . . many new industries that once were considered part of the periphery. . . . By contrast, *sunset industries* tend to include those that remain shackled by their past and have difficulties repositioning themselves vis-a-vis the new markets and the new economy. Many of these industries were once part of the core. It is almost as if there has been a complete reversal in the relationship between core and periphery, with the periphery comprising the services and high-tech industries now taking the lead.[59]

Noyelle goes on to argue that the new privileged position of sunrise industries enables them to deemphasize hierarchy and control and to emphasize flexible production and skilled workers with varying types of expertise.

It does indeed appear that the expert sector has increased in size in recent years. Census data reveals that class sectors shifted between

1960-1970 and 1970-1980: The rate of expansion of managerial and expert classes accelerated, and the numbers of working-class jobs in all employment sectors (but particularly in blue-collar production) were reduced.[60] It appears that the nature of capitalism is changing, incorporating new types of exploitation centered around new types of control: "Postcapitalist classes are rooted in control over two crucial kinds of productive resources—organizational assets and skill assets—both of which become increasingly salient as the forces of production develop."[61]

Another aspect of the changing nature of capitalism is the globalization of capital and the internationalization of production. As we have seen, nonexpert-sector jobs in diverse industries have been exported in recent years. It is therefore inappropriate to analyze class sector shifts in the United States only; "deproletarianization" at home is paralleled by proletarianization abroad. Moreover, even at home, deproletarianization does not necessarily imply upward class mobility at the level of individual workers; more commonly, workers are displaced into unemployment or low-level service sector jobs. Indeed, the jobs that are projected to grow the most between the present and 1995 include some high-tech occupations, but the highest rates of growth are predicted for such occupations as cashiers, janitors, truck drivers, waiters and waitresses, and food preparation workers.[62] Low-level service employment, unemployment, and deindustrialization render problematic the optimistic postindustrialist scenario of a highly educated leisure society.

A final way in which technocratic restructuring is affecting social stratification concerns race and sex segmentation. As we have seen, in diverse organizations, women and racial minorities are concentrated in the nonexpert sector and underrepresented in the expert sector. The erosion or abolition of internal job ladders has circumscribed mobility prospects for nonexpert workers and has fundamentally altered the shape of EEO issues. As Noyelle points out, whereas "previously focused on dismantling discrimination within internal labor markets, EEO policies must now pay much greater attention to the importance of the linkage between educational opportunities and employment opportunities."[63]

However, questions arise as to how sufficient educational credentials and technical expertise are for women and minorities in their quest for upward mobility. In a context of credential inflation, there is evidence of a shift from achievement back to ascription, underemployment, and differential returns to human capital investment for men and women.[64] As we have seen, women and minorities who achieve expert-level jobs face new types of discrimination and are usually found in the lower tiers of managerial or professional sectors:

> There is a rapidly developing split in professional work between prestige jobs with good pay, autonomy, and opportunities for growth and development and a new class of more routinized, poorly paid jobs with little autonomy and which are unconnected by promotion ladders to prestige jobs in the professions. . . . it is precisely in the newer, more routinized sector of professional employment that women's employment will be overwhelmingly concentrated.[65]

In an extensive study of sex segregation in a diverse sample of California firms, William Bielby and James Baron found "nearly complete" sex segregation, both among occupations and within organizations.[66] Even in occupations ostensibly desegregated, they found men and women having different job titles within the workplace; in their sample, over 96% of the women would have to be transferred to a different job title in order to equalize sex ratios.[67] Their statistical analysis reveals that employers' "taste for discrimination" is far more influential in creating sex segregation than are "supply-side factors" such as human capital differences. They conclude that "employers do reserve some jobs for men and others for women, based on their knowledge of technical and organizational features of work and their perceptions (however accurate) of sex differences in skills and work orientations."[68]

In a technocratic context of increased emphasis on expertise, technology, and instrumental rationality, gender stereotypes that define femininity as antithetical to these aspects of technocracy persist: Women are seen as subjective, emotional, irrational, unscientific, and technically incompetent.[69] Although physical strength is less important as a source of gender discrimination, new stereotypes have arisen.[70]

> How has hegemonic masculine ideology dealt with the shift of technology from heavy, dirty, and dangerous (electromechanical) technology to light, clean and safe (electronic) technology, given that masculinity was so clearly associated with the former qualities and femininity with the latter? The ideology has done a neat "about turn." The new technology is associated with logic and intellect and these in turn with men and masculinity. The complementarity is preserved by associating women and femininity with irrationality and physicality.[71]

Such stereotypes contribute both to keeping women segregated in the nonexpert sector and to female disadvantage within the expert sector. They also have ramifications for socialization and educational practices—for instance, in mathematics and computer training, which are correlated not only with sex but also with class and race.[72]

Gender and racial disadvantage is also apparent within the U.S. economy as a whole. Two-thirds of all adults living in poverty are women, more than half of all poor families are female headed, and nonwhite female-headed families are particularly disadvantaged.[73] Ironically, during a time of feminist movement and expanding awareness of women's issues, women's overall economic situation has deteriorated, as the occupational gains made by some educated women are offset by the deterioration in the economic status of the majority of women workers.[74] It appears that feminist gains at the level of ideology and political consciousness have been undermined by structural conditions that continue to operate against gender equality. Technocratic discrimination, which centers around polarization, sex segregation, and employer discrimination, perpetuates gender and race inequality despite the new options that are opened up during a time of changing organizational structure.

Finally, gender and race issues are salient within the international division of labor. Young, nonwhite women comprise 85% to 90% of workers in the export processing zones (EPZ) of U.S.-based multinationals.[75] Women are marginalized in newly industrializing societies, both by patriarchal oppression and by the transition from subsistence production to capitalism, which makes them particularly vulnerable to capitalist exploitation.[76] Managers of EPZ firms explicitly utilize both the ideology of technological/economic progress and traditional sexist stereotypes (beauty contests, cosmetics classes, and fashion shows) to promote worker docility and divert women workers.[77] The international division of labor reflects the extremes of technocratic polarization: young, nonwhite women performing work that is conceptualized and designed thousands of miles away by predominantly white male engineers.

TECHNOCRACY AND POLITICS

Although technocracy is typically presented as apolitical, purely rational, and beyond ideology, the evidence from technocratic workplaces indicates that this is far from the case. Technological design and technocratic restructuring have been shaped by political interests: capitalist interests, technical interests, military exigencies, and race and sex considerations.[78] Moreover, technocracy has political effects, some intended and some unintended, and new types of "technopolitics" are emerging as experts and managers forge new types of alliances, experience conflict, and use one another in diverse ways.[79] In this section, we explore technocratic politics not only at the level of the workplace, but also at societal and international levels.

One of the most striking societal reflections of technocratic politics is the emphasis on corporatist cooperation and collaboration among different types of experts.[80] In diverse advanced capitalist economies, trends toward labor/management cooperation, public-sector/private-sector interdependence, "codetermination," and interest-group intermediation and consensus are apparent. The underlying assumption of these corporatist trends is that of shared interests: that differences of opinion are minor and that "integration can be achieved through shared values, knowledge, and techniques."[81] Conflict is viewed as an irrational vestige that must be overcome in the interest of productivity and technical rationality.

In actuality, at the societal level as in the workplace, experts often disagree, and political conflict is transformed but not eliminated in a technocratic environment. Robert Putnam, in a comparative study of technocratic influences on government in Britain, West Germany, and Italy, found that although the "emergence of the hybrid figure of the politician-technician may be the most significant contemporary trend in elite composition," this development does not imply a diminished role for politics.[82] In fact he found substantial disagreement and differences in political orientation among various types of experts, with those technical experts trained in the natural sciences or technology more antipolitical and elitist in their orientation than social scientists. Although "technocrats agree that 'politics' should be replaced by 'rationality,' . . . on practical issues they may rarely agree which policy is uniquely 'rational.'"[83]

Another aspect of technocratic restructuring that affects national politics is the computerization of record keeping and administration. In the United States, computerization of government records has proceeded rapidly since the early 1960s, with 60% to 78% of the files of key agencies computerized by 1985.[84] Computer "matching" of records (comparison of two or more systems of records in order to check for fraud) has also become increasingly prevalent. Priscilla Regan has analyzed the political implications of these developments, concluding that the efficient management of government programs may well come into contradiction with individual rights and civil liberties, creating the possibility of unreasonable search and seizure suits.[85] Computerized crime data, a major feature of the technocratic administration of justice, also raises civil liberties issues. Such trends are more political than technical in nature, raising concerns about increases in state power and the erosion of democracy.

Within the international economic system, a similar technological capacity to promote centralized control and related dangers for developing countries has been noted. Advanced automation and the relocation of multinational production facilities to developing countries are no longer alternative policies but are increasingly occurring together. The choice of location of offshore plants will be increasingly selective, focusing on a few "industrial growth poles in Southeast Asia, Latin America, the Middle East and the Mediterranean area."[86] Moreover, pressures to rationalize the international production system have been increasing, particularly the perceived need to better coordinate and control geographically dispersed production activities.

> Thus, it has become possible to synchronize, on a worldwide scale, decentralized production with a strictly centralized control over strategic assets such as product design and plant layout; global cash management; logistic coordination; on-time operational control of production and complementary support services, particularly marketing and inventory. At the same time, global information networks present new possibilities for headquarters management to control affiliates around the world, to put them under pressure, if need be, and even to force them into a ruthless mutual competition.[87]

Such potential for increases in centralized control raises concerns about increased dependency and reduced autonomy of developing countries.

Some have pointed to the convergence between state-socialist and capitalist technocracy as a noteworthy trend, although interpretations of the implications and extent of this convergence have varied. Michael Burawoy contends that "all the evidence we have from state socialist societies suggests a striking similarity between their labour processes and those in capitalist societies."[88] Overall regulation of the economy differs, however, with capitalist societies relying more on market competition and socialist societies emphasizing centralized planning and plan bargaining between the central planning agency and enterprise directors. Neither system works particularly rationally, as "the anarchy of the capitalist market finds its analogue in the anarchy of the socialist plan."[89] Both systems have therefore been the target of technocratic rationalization efforts.

There are clear parallels between technocratic workplaces in capitalist and state-socialist countries. Socialist technocracy also generated a polarized workplace, what Burawoy terms a polarization into "core" workers and "peripheral" workers. Core workers are integral to the working of the firm, due to specialized knowledge or skill, and they are treated accordingly.

> The need to respond frequently and rapidly to changing requirements gives a great deal of power to the skilled and experienced workers, who over time develop a monopoly of knowledge essential to the running of the enterprise. From the management side the penetration of external uncertainties onto the shopfloor elicits two strategies. On the one hand management can seek to reward cooperation, particularly of the core workers; on the other it can intensify surveillance and control, particularly over the more peripheral workers.[90]

Moreover, Burawoy contends that the separation into conception and execution has been even more widespread under state socialism, with the conceivers or planners clearly differentiated from the executors, not only at the level of the workplace, but at the societal level as well.

As in capitalist technocracy, core workers in socialist technocracies are not designated as such solely on rational or meritocratic grounds. Political influences are also influential, in this case primarily the "macropolitics" of the party. However, micropolitics are also a factor. David Stark studied Hungarian "work partnerships" and found that the strategic workers who were rewarded differentially differed on "cultural capital" variables as well as human capital ones.[91] Educational credentials are apparently becoming more important in state socialist societies; a recent study of correlates of party membership in the Hungarian communist party found that "professionals and technocrats are the most likely candidates for membership in the MSZMP."[92]

Other recent studies of socialist technocracy have focused on the more general role of technocrats in both the production system and the political system. Tom Baylis, in an intensive study of technical experts in governmental and production sectors in East Germany, analyzed the extent and nature of technocratic influence.[93] He found the technical intelligentsia to be a new elite, with their technical expertise closely allied with political motives.

> The policy influence of the technical strategic elite can be said to be substantial, established, and partly institutionalized, but still uneven in its effectiveness and often divided in its thrust. The most important and effective of its members are not "technocrats" in the sense of being apolitical professionals who have somehow acquired governmental power, but politician-specialists with economic training and technical interests who have learned to adapt to a rapidly changing political environment and sometimes to manipulate it on behalf of their preferred goals. The direction of their influence is not uniform because they differ in the weight each gives to his "political" and "professional" sides and because even in their professional judgment they do not entirely agree on the measures most likely to maximize economic rationality.[94]

In general, Baylis found East German politics to have changed in style but not in substance, and the technical elite and the *apparatchiki* to be "partners in power rather than rivals for it."[95]

Other analyses of socialist technocracy have seen more far-reaching effects of recent transformations. Mallet, for instance, saw technocratic rationality as a progressive force, prefiguring a more democratic socialism by coming into contradiction with prevailing bureaucratic power.[96] In a similar but more sustained analysis, Bahro analyzed both the contradictions of "bureaucratic/centralist" organization and the growing role of technical experts.[97] Bahro saw the "planning from above rationality" as limited by a lack of knowledge of social needs ("from below") as well as by the fact that it cannot be objective and scientific in its planning while operating in a context of a stringent and unequal division of labor. Competence at every level is overly circumscribed.

Bahro placed great hope in the growth of knowledge and expertise, for they generate what he terms "surplus consciousness" and "surplus expertise," capacities that cannot be utilized fully within a bureaucratic division of labor. Expertise in any field generates a capacity for abstraction and reflection that implies a capacity to be engaged in political decision making and hence potential challenges to domination.[98] Bahro therefore sees technocratic rationality as inherently universalistic and democratic and as challenging the vested interests of the bureaucratic status quo. Although scientism and technical rationality are not in themselves sufficient, they prefigure a fuller definition of progress and a "new social revolution."

George Konrad and Ivan Szelenyi see the emerging technocratic elite in Eastern Europe as occupying a more contradictory position, as both allied with the ruling elite and as challenging this elite.[99] Konrad and Szelenyi differentiate between the technical elite, which is increasingly integral to production and redistributive planning, and the more humanistic intelligentsia, which is increasingly marginalized. Although the technical elite is tied to the status quo, it is also structurally antagonistic to it.

> The technocracy has not been content merely to see its material and social status rise; by vindicating professional knowledge and achievement as principles which legitimate power, it has called into question the power monopoly of the ruling elite. Nor can it demand that expert knowledge and performance bring recognition for the technocrats alone. In this respect they are compelled, like it or not, to represent the interests of the whole intellectual class. No technocrat can feel secure so long as the best minds in science and art languish in prison or are forced to earn their living at menial jobs which stultify their creative powers.[100]

Like Bahro, Konrad and Szelenyi see knowledge and expertise as embodying a progressive dynamic, offsetting any elitist tendencies, and as prefiguring a "new stage of socialism." However, they argue that these progressive tendencies can only be realized if the technical elite allies with the marginal intelligentsia and the working class.

The dramatic and revolutionary events of 1989 in Eastern Europe and the former Soviet Union have highlighted the role of the intelligentsia. In Czechoslovakia, playwright Vaclav Havel and a group of independent intellectuals assumed power and effected a peaceful transition. The intelligentsia have also played a prominent role in Yugoslavia, East Germany, Hungary, and Poland, although in the case of Poland the Solidarity movement was based upon an alliance of workers and intellectuals.[101]

In the Soviet Union as well, the technical intelligentsia were deeply involved in Gorbachev's *perestroika*.

> Gorbachev has gone to great lengths to involve the Soviet intelligentsia in making *perestroika* work. Yet as the situation advances past the "less" stage (less Party control over spheres of life, less KGB involvement in domestic affairs, less violation of the law by officials, less control over

the press), the intelligentsia may be expected to come up with more sophisticated and "rational" proposals for reform. . . . The avowed aim of "getting the Party out of the details of daily life" means that decisions will increasingly be taken at the level at which they will have to be implemented. The purpose of this is clearly to make those decisions more closely reflect the real conditions and the real demands of the situation, and not the needs or interests of particular groups or individuals. Thus, decisions are to be made more and more on technical grounds, on the basis of the recommendations of "experts."[102]

Thus, one of the key problems that *perestroika* was designed to address was the technical backwardness of Soviet science and the limitations created by the bureaucratization of such professions as medicine, law, journalism, and engineering, and both the technical and the humanistic intelligentsia have been at the forefront of recent reform efforts. Given the "modern scientific, technological, administrative and intellectual tasks of the reform"[103] in the former Soviet Union, technical experts have been allied with reformist state officials from the beginning.

However, the failed coup attempt of August 1991 in the former Soviet Union and the subsequent restructuring of the Soviet Union have revealed the limitations of such elitist restructuring. Although the reforms were initiated by certain state officials in conjunction with the intelligentsia, it was the more grassroots constituency of Boris Yeltsin that was necessary to preserve and extend the reforms. Although the process of change is far from over in the former Soviet Union, it appears that political coalitions of workers and intellectuals will be at the forefront of future changes.

Such a coalition of workers and professionals was also apparent in the Solidarity movement in Poland. Solidarity, however, emphasized the political importance of workers to a greater extent than other reform movements in Eastern Europe.

Polish engineers and physicians sided with Solidarity. . . . Engineers could not make the economy more rational, nor increase their own professional autonomy from the authorities, without the alliance with workers. This was especially obvious in the move toward self-management. Engineers were among the most active in this movement and they were the ones who would be elected to new managerial positions. But this self-management was not a technocratic dream, because engineers could lead only in so far as they could convince workers, through enterprise elections, of their right to lead. Self-management is the perfect expression of this class alliance.[104]

Michael Kennedy argues that the centrality of workers in Solidarity has led to a reform movement where social equality is as important as economic rationality, whereas the more elitist movements in the former Soviet Union and Hungary (where intellectuals and students have been the vanguard) have led to both moral and practical political problems: Workers presumably stand to benefit from the reforms, but they are less identified and involved with the changes.[105] Of course, the most recent upheavals in the former Soviet Union may change the class alliances and give the working class more direct involvement in subsequent changes. The politics of the working class are not inherently universalist or progressive, however; right-wing nationalism and anti-Semitism have been associated with the Polish working class movement, for instance.[106]

It appears, then, that both theory and practice within Eastern Europe and the former Soviet Union point to the intelligentsia as a "flawed universal class," as both "the center of whatever human emancipation is possible in the foreseeable future" and as a possible new elite.[107] What has received less attention from neo-Marxist scholars is the fact that the political perspective of the working class may also be limited in certain respects. Clearly, in both capitalist and state socialist countries, politics has not been rendered obsolete by technocratic developments, but has been made more important. Only through the old-fashioned politics of coalition building, resistance, and conflict can the universalistic potential of technocracy be realized.

Conclusion

There is, then, a new type of technocratic work organization emerging in diverse occupational settings. Both the structure and the ideology of workplaces have changed in recent decades, as technological innovation has opened up opportunities for social and political restructuring, opportunities which have only been partially realized.

Technocratic restructuring, while not monolithic, is assuming generalizable form. The basic characteristics of technocracy are: (1) Polarization into expert and nonexpert sectors, with corresponding race and sex segregation and dichotomous working conditions; (2) Varying configurations of centralization/decentralization, with the norm being underlying centralization of control masked by a more superficial type of decentralization; (3) Skill restructuring, with more abstract, diagnostic, and technical skill displacing more traditional skills; (4) An emphasis on demonstrated expertise and credentialism rather than the rank authority of bureaucracy; (5) A new type of technocratic ideology which alleges that technical imperatives have displaced traditional politics, although in actuality new types of technocratic politics are operative.

The social implications of emergent technocratic restructuring are considerable. It appears that technocratic polarization at the level of the workplace is paralleled by a corresponding polarization of the class structure, at both national and international levels. Growth of the expert sector is offset by a corresponding increase in low-level service-sector employment, unemployment, and underemployment. Women and racial minorities are particularly disadvantaged by technocratic restructuring, in that they tend to be overrepresented in the nonexpert sector and to face new types of discrimination.

Technocratic politics involves complex patterns of interaction among different types of experts, managers, and politicians. In socialist and capitalist countries alike, the political role of technical experts has expanded, although the technocratic ideology of apolitical decision making based solely on technical criteria has not been realized. Rather, the emerging technocratic elite coexists with existing political and economic elites, at times in coalition with them, at times in confrontation. Whether technocratic restructuring will promote progressive social change in the future will depend upon the political alliances which are made, and in particular whether non-elite political groups and grass-roots constituencies have a voice in political decision making. The final chapter deals more fully with such political questions and possible scenarios of twenty-first century society.

Chapter 7

CONCLUSION

The irrationality of domination, which today has become a collective peril to life, could be mastered only by the development of a political decision-making process tied to the principle of general discussion free from domination. Our only hope for the rationalization of the power structure lies in conditions that favor political power for thought developing through dialogue. The redeeming power of reflection cannot be supplanted by the extension of technically exploitable knowledge.
—Jurgen Habermas, "Technical Progress
and the Social Life-World"

This is a critical juncture in the development of work organizations, as well as for societies in general. Decisions that are being made now will affect the working lives of men and women well into the twenty-first century. In this chapter, I will situate technocracy within a broader political context, so as to illuminate future probabilities, possibilities, and dangers. Implicit is the view that people make their own history, but that they do not make it just as they please, and hence that understanding past and present realities is as important as political vision. The previous chapters have focused on understanding past and present realities; this chapter attempts to use this knowledge to forge a realistic political vision. I will begin by examining others' conceptions of the politics of changing work organizations and then proceed to my own analysis of how the emergence of technocratic organization opens up new possibilities for reform and further restructuring.

CONTEMPORARY VISIONS OF SOCIOECONOMIC CHANGE

As we have seen, one pronounced trend in existing social-scientific and journalistic analyses of changing work organizations is their emphasis on the positive effects of computerization. The main reason for the strongly optimistic tone of many recent analyses of computerization is the fact that they focus on the expert sector (either exclusively or primarily). The literatures on "postbureaucratic organizations," "postindustrial organizations," "postmodern organizations," and "adhocracies," for instance, all emphasize the type of working conditions characteristic of the expert sector: diminished hierarchy and control, de-differentiation, horizontal rather than vertical integration, team organization, extraorganizational networking, informalism, upskilling of work, increased prospects for democracy.[1] This trend has also been pronounced in the popular press, where expert-sector work has been taken as paradigmatic for working with computerized technology.

Conversely, those analyses that have been more critical of the ways in which computerization has been implemented have tended to derive from empirical studies of nonexpert-sector work. Emphasized in these accounts is the way in which the computerized system can be used to enhance supervision and control: that nonexpert-sector workers suffer from reduced autonomy, deskilling, and new types of stress and occupational health problems. As with the optimistic scenarios, these analyses are not so much wrong as they are partial.

Even those theorists who have been aware of a greater duality inherent in computerization have not often made clear the fact that this dualism is usually present within individual firms; Zuboff, for instance, fails to discuss automating and informating as twin strategies that are often used for different sectors of the same workplace, and Kanter generally discusses integrative and segmentalist organizations as two distinct types of organizations, rather than as different features found within the same workplace.[2] What is not being adequately theorized is the bifurcation into expert and nonexpert sectors, which is one of the central characteristics of technocratic organizations.

Social scientists who have attempted to envision macrosocial political and economic alternatives have also been limited by their inattention to the new class divisions within workplaces. Piore and Sabel, for instance, see computerization as promoting a more craftlike orientation to work, one in which the machine "responds to and extends the productive capacities of the user."[3] Given the empirical findings on computerization, it seems clear that the main thrust of their analysis is on the "community of the skilled," the expert sector.

More importantly, what Piore and Sabel fail to adequately theorize is that Fordist techniques and flexible specialization currently coexist in technocratic organizations. They do argue that the two types of organization could be compatible, but within the international division of labor: that "the old mass-production techniques might migrate to the underdeveloped world, leaving behind in the industrialized world the high-tech industries . . . all revitalized through the fusion of traditional skills and high technologies."[4] To endorse such a dualism within the world economy seems questionable, but their inattention to the corresponding duality within work organizations renders their entire political scenario problematic. The achievement of "yeoman democracy," while a laudable political goal, is more difficult in a context of pronounced polarization.

Robert Reich does recognize that bifurcation and polarization are dangerous and pronounced trends within the global economy.[5] In his terminology, the symbolic analysts are virtually seceding from routine production workers and in-person service workers: In terms of material rewards, working conditions, living conditions, and general societal power, one can argue that they have already achieved secession. For Reich, the main question becomes how to stem this trend toward secession, how to achieve the national political will to increase public investment and raise the standard of living of nonexpert-sector workers. Economically, nationalism makes little sense, due to the complex production and ownership webs of the global economy. Reich argues that a new type of nationalism is needed:

> . . . a positive economic nationalism, in which each nation's citizens take primary responsibility for enhancing the capacities of their countrymen for full and productive lives, but who also work with other nations to ensure that these improvements do not come at others' expense. This position . . . rests on a sense of national purpose—of principled historic and cultural connection to a common political endeavor. It seeks to encourage new learning within the nation, to smooth the transition of the labor force from older industries, to educate and train the nation's workers, to improve the nation's infrastructure, and to create international rules of fair play for accomplishing all these things.[6]

However, although Reich does recognize that the living conditions of nonexpert-sector workers should be improved, and has practical ideas about how to curb the trend toward the secession of the experts, his analysis is still biased in favor of the expert sector. Clearly he sees symbolic analysts as the ascendant class in the contemporary world economy, and the main goal for the United States as "helping Americans become technologically sophisticated."[7] The main thrust of his policy recommendations is to reform capitalism so as to achieve more equal opportunity—enhanced opportunity to become a symbolic analyst.[8]

Another problem with Reich's analysis is his overemphasis on the decentralization of the world economy and the "diffusion of ownership and control."[9] Certainly there is some truth in his analysis of the decentralized web that the global economy has become, as well as the way in which knowledge and specialized expertise have become important new sources of power. When more than 85% of the price of a semiconductor chip is for design, engineering, and patents and copyrights of past design, only 6% is for routine labor, and 8% is profit, clearly the structure of capitalism and surplus value extraction are changing.[10] However, the capitalist system, now worldwide in scope, does seem to be surviving. Richard Brahm and Marc Jones, for instance, argue that new forms of capitalist control of the decentralized production webs are emerging, pointing toward a more powerful consolidation of multi-industry oligopoly structures that transcends the nation-state.[11] As with the apparent trend toward decentralization of technocratic workplaces, at the level of the world economy as well there is a danger of overemphasizing the superficial level of decentralization and ignoring the hidden but powerful forces that maintain centralized control.

A thorough and detailed vision of how to begin to restructure the U.S. economy in a more democratic direction is provided by Block.[12] He not only analyzes the importance of workplace democracy, the fact that employee participation is not necessarily linked with employee ownership, and the potential contribution that workplace democracy might make to productivity and innovation, but he also goes on to discuss the broader socioeconomic changes that would be necessary to make workplace democracy feasible. These include both changes in the organization of the workplace (e.g., meaningful job ladders and career development opportunities, greater employment security, protection of free speech, employee rights to own stock and consent to takeovers, etc.) and changes in the overall organization of the labor market: basic income supports for all members of society (working and nonworking), increases in voluntary leisure time, expansion of nonmarket work, expansion of part-time work opportunities, an "expanded public and non-profit domain of knowledge production that extends its research scope to production processes,"[13] and new combinations of markets, state regulation, and other types of regulation. Block's vision is well conceived, and it merits attention. However, his inattention to the obstacle that expert/nonexpert polarization represents, as well as his overly optimistic evaluation of the impact of computerization on the workplace, undermines the utility of his vision.

Other theorists of organizational alternatives have turned toward other countries as exemplars of progressive organizational change. Japan, for instance, has gotten a lot of attention, being praised for its economic success, its "quality circles," and its "flexible-system production."[14] Stewart Clegg, in fact, goes so far as to see Japanese workplaces as harbingers of postmodern organization, which he defines in the following manner:

> Where modernist organization was rigid, postmodern organization is flexible. Where modernist consumption was premised on mass forms, postmodernist consumption is premised on niches. Where modernist organization was premised on technological determinism, postmodernist organization is premised on technological choices made possible through "de-dedicated" microelectronic equipment. Where modernist organization and jobs were highly differentiated, demarcated and de-skilled, postmodernist organization and jobs are highly de-differentiated, de-demarcated, and multi-skilled.[15]

However, as Clegg notes, although Japanese workplace innovation has been considerable, significantly going beyond Fordist and bureaucratic models (what Clegg calls "modernist" organizational forms) and achieving a high level of efficiency and economic competitiveness, as a model of progressive change it is limited in certain respects. The overall occupational structure is hierarchical and privately owned, rendering Japanese workplaces quasi-democratic (or pseudodemocratic) at best. Moreover, the Japanese labor market is highly polarized between predominantly male workers in the core labor market and the more than two-thirds of all workers, disproportionately women, who are peripheral to Japanese postmodernity and excluded from its benefits.[16] The Japanese system appears to exhibit a polarization similar to technocratic organizations in the United States, with expert-sector workers privileged at the expense of nonexpert workers.

For a more democratic and progressive model of postmodernism,

Clegg looks to Sweden. In the Swedish economy, collective capital formation has evolved to the extent that a higher degree of economic democracy has been achieved, and indeed the capital/labor distinction has been
blurred: another model of postmodernist "de-differentiation." Since the
1920s, Sweden has been concerned with promoting both workplace democracy *and* economic democracy, with the Swedish Social Democratic Party
putting many of these ideas into practice.[17] One policy of particular relevance to the present argument was a wage policy that emphasized wage
increases for the lowest paid workers and wage restraint among highly
paid workers in core sectors of the economy—a policy that resulted in high
profits by the mid-1970s.[18] In recent years, the emphasis has primarily been
on putting these profits to use by the development of wage earners' funds
and a gradual shift of ownership and control to workers—the famous Mei-
dner proposals, adopted in partial and limited form in 1984.

Social scientists who have attempted to conceptualize organizational alternatives, then, have drawn inspiration both from the past
(e.g., from craft organization) and from cross-national examples. While
instructive, these analyses have generally failed to adequately address
the polarization associated with technocratic restructuring, and inattention to this polarization has led to partial approaches that adopt the
stance of either the expert or the nonexpert sector, failing to comprehend the full political implications of technocratic restructuring.

POSTMODERNISM, LATE CAPITALISM, AND TECHNOCRATIC CHANGE

As we saw in chapter 2, the growing influence of science and
expertise has inspired many diverse theoretical commentaries, from
the Enlightenment faith in, and advocacy of, technical experts to the
deep skepticism concerning scientific rationality characteristic of poststructuralism and postmodernism. What kinds of political practice and
social alternatives are offered by the postmodernist approach?

Postmodernism is a diffuse and primarily cultural phenomenon,
although some have linked it with changes in work organizations.
Clegg, for instance, emphasizes *de-differentiation* as the central characteristic of postmodern organizations.

If an analysis of the management of modernity appears now to be relatively unproblematic, what of postmodernity? If the key feature of this

process is the de-differentiation and disassembling of extant forms of the division of labor then this is tantamount to a deconstruction of some, at least, of the bases of modern management. Whatever postmodern management might be, it is unlikely that it would be based on the same organizational practices of differentiation as had been modern management.[19]

As we have seen, one of the central aspects of technocracy is the erosion of the hierarchical division of labor characteristic of bureaucracy and the emergence new types of technocratic control. De-differentiation, then, does not necessarily imply the "deconstruction" of management, but rather its transformation.

A broader and more comprehensive conceptualization of postmodernism is provided by Jameson.[20] For Jameson, postmodernism is the "cultural logic" of late capitalism, and, following Ernest Mandel, he sees late capitalism, or multinational capitalism, as the third and most recent stage of capitalist development.[21] Technocratic innovation is certainly related to the broader evolution of the entire capitalist system, including its cultural dimensions. Indeed, as Jameson points out, what postmodernism represents is a "cultural revolution" at many levels, including that of the organization of production.[22]

Postmodernism has been invoked so often, in so many different contexts, that it has become rather conceptually elusive. For Jameson, postmodernism is primarily a cultural development, but one with direct links to political economy and changes in work organizations. Indeed, for Jameson, one distinguishing feature of the postmodern age is that fact that the "semiautonomy" of the cultural sphere has been destroyed. Postmodern architecture, for instance, with its bewildering mazes, is for Jameson a reflection of a more general political disorientation.

> This latest mutation in space—postmodern hyperspace—has finally succeeded in transcending the capacities of the individual human body to locate itself, to organize its immediate surroundings perceptually, and cognitively to map its position in a mappable external world. It may now be suggested that this alarming disjunction point between the body and its built environment—which is to the initial bewilderment of the older modernism as the velocities of spacecraft to those of the automobile— can itself stand as the symbol and analogon of that even sharper dilemma which is the incapacity of our minds, at least at present, to map the great global multinational and decentered communicational network in which we find ourselves caught as individual subjects.[23]

Other characteristics of postmodernism for Jameson include an erosion of the distinction between "high" and "low" culture, a new "depthlessness," a "weakening of historicity," and a "whole new type of emotional ground tone" or "intensities," with all of these closely related to advanced technology and the new economic world system.[24]

As we have seen, Lyotard has defined postmodernism in opposition to "metanarratives," particularly those with their origins in the Enlightenment, due to their alleged conservativism.[25] The argument here is that the attempt to rationally understand the totality points toward totalitarianism, and that the Enlightenment emphasis on reason and rationality is suspect. Lyotard's critique of instrumental rationality, "performativity," and the ways in which computerization extends and magnifies the threats to freedom are incisive (if not terribly original). However, his proffered alternatives of narratives, diverse language games, and paralogical thought seem limited as political alternatives, however appealing they may be as intellectual exercises.

What are we to make of this strange new landscape of postmodernism? Although this literature contains important critiques of contemporary reality, the politics of postmodernity is flawed. Theoretical challenges to the binary oppositions of modernism will not cause them to disappear, and empirical evidence indicates that polarization continues to be apparent at many different levels. The postmodern emphasis on pluralism, complexity, and decentralization can obscure (rather than challenge) existing forms of domination, and the postmodernist skepticism regarding modernist rationality can impede the development of a coherent political vision that might begin to realize the democratic reality postmodernists endorse. Postmodernist art and architecture reflect the complexity of the modern world system in emotionally appealing ways: As Jameson points out, the architecture creates the feeling of wandering in a confusing maze while playfully enjoying the experience. However, playing in the maze is more an existential stance than a viable politics.

Marginalized groups, such as women, racial minorities and homosexuals, have tended to agree with the postmodern critique of Englightenment rationality, arguing, for instance, that "transcendental claims reflect and reify the experience of a few persons—mostly white, Western males."[26] However, Seyla Benhabib defends the tradition of critical theory vis-à-vis postmodernism.[27] She argues that critical theory has been clearer with regard to its "cognitive and moral criteria," defining these as "the defense of a communicative, discursive concept of reason, the acceptance that knowledge should serve moral autonomy, and the recognition that intentions of the good life cannot be dissociated from the discursive practice of seeking understanding among equals in a process of communication free from domination."[28] Conversely, Benhabib finds Lyotard endorsing a naive type of pluralism, one that neglects structural sources of inequality:

> In the absence of radical, democratic measures redressing economic, social, and cultural inequalities and forms of subordination, the pluralistic vision of groups Lyotard proposes remains naive. It would fail to redress the plight of those for whom the question of the democratization of information is a luxury, simply because, as marginalized groups in our societies, they fail even to have access to organizational let alone informational resources. At the present, these groups include increasing numbers of women, minorities, foreigners, unemployed youth, and the elderly.[29]

Postmodernist pluralism, then, is a naive hope, at best, and an ideological facade masking political indifference, at worst.

The tradition of critical theory presents an alternative critique of instrumental rationality. The Frankfurt school, in such works as *The Dialectic of the Enlightenment* and *One-Dimensional Man*, incisively criticized the growing dominance of instrumental rationality, although without offering much sense of a viable alternative.[30]

More recently, Jurgen Habermas has presented a more optimistic alternative, one that criticizes instrumental rationality but defends an expanded conception of reason. For Habermas, postmodernism is a conservative flight from the "project of modernity" and its related Enlightenment values, particularly the idea of a type of rationality that would include both moral/practical and aesthetic/expressive dimensions.[31] Habermas has also challenged the position that science must be rejected in favor of new approaches to knowing, arguing instead that scientific and technical rationality have their place and become dangerous only when they present themselves as the only valid approaches to knowing.[32] Another major contribution to the critique of technocracy has been Habermas's analysis of the legitimation and motivation crises that can plague social systems, an analysis that used systems analysis against itself.[33]

In the confrontation between Habermas and the postmodernists, I believe that Habermas has more to offer in the way of political vision, for Habermas has done more to promote the realization of democracy and pluralism and has been more judicious in his critique of instrumental rationality. With his analysis of the demise of the public sphere and "systematically distorted communication," and his vision, however imperfect and preliminary, of an "ideal speech situation" in which the only power would be the force of the better argument, Habermas has at least pointed toward some of the issues that need to be addressed if technocracy is to be democratized.[34] I turn now to a closer examination of how an understanding of the nature of technocratic organization might inform and further elaborate such visions: how existing contradictions may open up realistic possibilities for change in contemporary workplaces.

THE FUTURE OF TECHNOCRACY

Like those who have extolled the capacity of advanced technology to open up opportunities for more interesting work and a more democratic workplace, I also see such potential. Unfortunately, however, empirical reality indicates that such potential has not been widely realized. The general pattern has been for workplaces to develop an increasingly polarized structure, with the expert sector reaping the benefits of "informated" work, enhanced autonomy, and collegial working relations, and the nonexpert sector continuing to work in a more traditional context, with computerization implemented so as to promote automation, routinization, and enhanced supervision. As we have seen, such polarization has tended to be legitimated by a technocratic ideology of technical imperatives and apolitical decision making.

However, this elitist model of workplace organization is not a seamless fabric. The rather extreme polarization within the same firm, highlighted by race and gender disproportionality, implies visibility and potential volatility: With largely stagnant internal labor markets, and hence little opportunity for upward mobility, nonexpert workers may begin to demand some of the advantageous working conditions that they see expert-sector workers enjoying (and hear about from the mass media as characteristic of "working with computers"). When the firm's official cultural ideology proclaims that all workers are part of "one big family," nonexpert-sector workers may begin to demand more egalitarian working conditions for all family members. Moreover, management may begin to realize that innovative ideas can come from all levels, and that nonexpert workers *are* as important to the functioning of the organization as expert workers. When expert workers are segregated from nonexpert workers, the ensuing lack of communication implies diverse problems in coordination of the work process—for instance, the inability to coordinate design and implementation.[35]

Democratization, then, must begin with the erosion of polarization and the equalization of working conditions. A revival of largely stagnant internal labor markets, and the creation of learning opportunities and job ladders for all workers, should be part of this effort. Income differentials should also be reduced, and in this regard the United States could perhaps learn from Sweden's example. The intrinsic rewards of expert-sector work could make the increased inequality of earnings characteristic of high-tech workplaces unnecessary (if not dysfunctional).

The Swedish welfare state might also prove instructive in dealing with the problem of unemployment. Given the potential of super-automation to create technological displacement of workers, both income support policies and an enlarged role for the state in job creation are necessary. Although this book has dealt primarily with employed workers, social problems related to the growing class of the unemployed are likely to be prominent in the twenty-first century.

Another aspect of organizational democratization might be an expanded usage of teams and task forces. Many types of nonexpert work could be effectively organized around task forces, job rotation, and the like. Moreover, teams that combine expert and nonexpert workers would have advantages in terms of integration of the labor process and promoting learning and skill enhancement, although teams of workers of varying statuses would need to be sensitive to the micropolitics of polarization and discrimination.

As Fischer has pointed out, what is needed are more democratic models of participatory expertise.[36] Such models, which have taken their inspiration from some of the new social movements of recent decades, transform the experts' role into one of *facilitating* dialogue and empowerment. The professional/client relationship, for instance, becomes one where more democratic exchanges and mutual learning can occur. Social-scientific research methodology similarly becomes more phenomenological in nature, emphasizing actors' "ordinary knowledge," joint learning, and dialogue. The goal of these efforts is to bring together the more formal, abstract knowledge of experts and the more informal, contextual, and often narrative knowledge of nonexperts in a "mutually beneficial problem-oriented dialogue."[37] Experts become analogous to teachers, facilitating a societal learning process that will define problems, clarify values and goals, and "bring theoretical and empirical knowledge to bear on the participants' circumstances, [although] it is the participants themselves who must actually decide which courses of action they are willing to undertake."[38]

The decentralization of technocratic workplaces has the potential to open up such participatory and democratizing practices. We have seen how microcomputers can be used to facilitate worker programming as well as communication links between relatively autonomous "satellite stations" and work teams. However, in practice these potential benefits of decentralization have been more common in the expert sector. In the nonexpert sector, the more common form of implementation has utilized a superficial and visible type of decentralization of the hardware, combined with an underlying, centralized control programmed into the design of the system. Once again, the more genuine decentralization found within the expert sector could serve as a model for a more thorough decentralization and democratization of the entire organization. Managers may well discover that the more centralized and authoritarian model creates more problems of control than it solves, and that nonexpert-sector workers would respond favorably to more opportunities for autonomy and responsibility.

Perhaps the most important aspect of the restructuring of technocratic organizations is that the design and implementation of new technological systems need to become more democratic processes, with social and political factors receiving as much attention as technical ones. In order for such a possibility to become realized, several related dimensions of technocratic ideology must be challenged: the idea that technology develops according to its own autonomous and exclusively technical logic of development, that technological imperatives determine the shape of workplace reorganization, that there is one best technical solution to any problem, and that such technical solutions can only be found by technical experts, presumably acting in the general interest.

Technocratic ideology is well entrenched and pervasive, and the repoliticization of technological development and implementation will therefore be a slow and difficult process. Such repoliticization needs to occur at many different levels: local, national, and international. However, one important starting point could be individual workplaces; if nonexpert-sector workers can become more involved and knowledgeable concerning technological systems, they will be in a better position to challenge technocratic ideology and participate in the political process of designing alternatives to technocratic organizations, alternatives that will be more genuinely represent the general interest. Hopefully this politicization will snowball so as to challenge technocratic decision making in different contexts and at national and international levels as well.[39] As people come to realize that there is no one best technical solution to problems in their workplace, they may also become aware that national and international political realities are similarly complex and require an expanded and transformed type of democratic political practice.

If the positive potential of advanced technology is to be realized, technocratic organizations must be democratized and technocratic rationality must be replaced by a more substantive political rationality that considers long-term implications and much broader social and political parameters. In this long and difficult process of societal restructuring, there can be no substitute for the politics of coalition building, debate, and negotiation among various constituencies, including, in particular, significant numbers of women, racial minorities, and nonexpert workers of various types. Only in this manner can we hope to achieve "thought developing through dialogue"[40] and the ability to successfully challenge technocratic organization and ideology.

NOTES

CHAPTER 1

1. On the concept of organizational control structure, see Etzioni 1965 and Heydebrand 1979, 1983.

2. My use of the term *technocracy* is derived from the work of Wolf Heydebrand, my mentor in graduate school. See Heydebrand 1979, 1983, 1989; Burris and Heydebrand 1981. Heydebrand's own work differs from mine in focus, largely because he gives less emphasis to advanced technology as a causal variable. In recent years he has focused more on empirical exploration of technocratic changes in the judiciary (Heydebrand and Seron 1990) and has abandoned the concept of technocracy in favor of "technarchy" (see Heydebrand 1985).

3. In using the terms *expert* and *nonexpert*, I do not mean to imply that this distinction necessarily corresponds to actual levels of expertise. In fact, the evidence indicates that many nonexpert workers have expertise that they are not able to use on the job, and that expert-sector workers may have a narrow and specialized type of knowledge rather than general expertise. The terminology of *expert* and *nonexpert* is rather meant to imply the managerial, and indeed general societal, definition of these sectors as such.

4. See Collins 1979, 1-2, for a perceptive caricature of this version of the technocracy argument.

5. See Collins 1979, chap. 1.

6. See Kouzmin 1980 for a good defense of looking intensively at technology as an important organizational variable.

7. Urban 1978, 54.

8. See Heydebrand 1983 and Burris and Heydebrand 1981 for a fuller discussion of the process of dialectical rationalization.

9. Bendix 1956, 205.

10. Weber 1978, Ulrich 1982.

11. Mill 1848, cited in Bendix 1956, 47.

12. Braverman 1974, Reckman 1979.

13. See Ulrich 1982 for a fuller discussion. Women were routinely expected to be "deputy husbands," performing male tasks, when their husbands were temporarily or permanently gone. This flexibility led to women's being represented in a wide range of trades and occupational activities during the colonial period, but within the context of patriarchal control and limited legal rights.

14. Although this book focuses on capitalist technocracy, there have been analogous developments in state socialist countries. See chapter 6 for a comparative discussion. On the demise of precapitalist forms of organization, see Clawson 1980 and Marglin 1974.

15. See Clawson 1980, Staples 1987, Storey 1983. The putting-out system allowed for home manufacture within the context of mercantilism and emerging capitalist appropriation of the product. Internal subcontracting, which was widespread in industries such as steel, machining, and coal throughout the nineteenth century, retained certain aspects of craft organization (e.g., personalized work groups, decentralized authority, work performed by the authority figure) but within the context of the capitalist factory. Internal contractors, who often hired their own family members, relied heavily on simple control and had virtually complete authority over their employees, although wages and prices were set by the capitalist owner. Internal contracting was thus an intermediate and contradictory stage between craft organization and capitalist control.

16. Thompson 1963.

17. Edwards 1979.

18. Storey 1983.

19. Storey 1983, 104.

20. As Bendix 1956, 115, points out, this entrepreneurial ideology represented a marked shift from traditionalism:

This new entrepreneurial ideology . . . had major significance in the development of English industrial society. For . . . a new formula had been found on the basis of which employers and workers were conceived as members of the same community. Evangelism had continued the traditional belief in the responsibility of the rich and dependence of the poor, a belief which affirmed the deep division between classes but which also asserted their interdependence. Now, the doctine of self-help proclaimed that employers and workers were alike in self-dependence, and that regardless of class each man's success was a proof of himself and a contribution to the common wealth. There was evangelical zeal in this appeal of employers to the drive and ambition of the people. By bidding the people to seek success as they did themselves, the employers manifested their abiding belief in the existence of a moral community regardless of class."

This new and more individualistic sense of moral community thus relied on the ideology of equality of opportunity. Both simple control and its corresponding ideology of individual ability and rights were also highly personalized, granting wide discretion to individuals assumed to be superior and worthy. Such personalized ideology is also vulnerable to refutation based on personal criteria: Particular individuals might be judged "unworthy" based on their behavior or perceived ability.

21. See Edwards 1979, Beniger 1986.

22. Edwards 1979, 110.

23. See Edwards 1979, Stone 1974.

24. Gouldner 1976, 182.

25. Montgomery 1979, 2.

26. Stone 1974.

27. Edwards 1979, Stone 1974, Montgomery 1979.

28. See Stone 1974.

29. Beniger 1986, 6.

30. See Weber 1978.

31. As Beniger (1986, 15) puts it: "The reason why people can be governed more readily *qua* things is that the amount of information about them that needs to be processed is thereby greatly reduced and hence the degree of control . . . is greatly enhanced. By means of rationalization, therefore, it is possible

to maintain large-scale, complex social systems that would be overwhelmed by a rising tide of information they could not process were it necessary to govern by particularistic considerations . . . that characterize preindustrial societies."

32. See Bendix 1956, Weber 1978.

33. See Bendix 1956, 232.

34. Alvesson 1987, 160.

35. Fischer 1984, 175.

36. Bendix 1956.

37. Bendix 1956, 240; see Scott 1966 for a good review of the literature on professional/bureaucratic conflict.

38. Freidson 1984, Wilensky 1964, Larson 1977.

39. Wilensky 1964.

40. Daniels 1973.

41. Freidson 1984, Wilensky 1964.

42. Larson 1977, Wilensky 1964.

43. Freidson 1973, Larson 1977.

44. Larson 1984, 34.

45. Larson 1984, 53.

46. Daniels 1973, Collins 1979, Haskell 1984a.

47. Benson 1973, Scott 1966.

48. Braverman 1974.

49. For instance, the type of steel used for machining was improved; see Clawson 1980, 243 and passim.

50. Stark 1980, Clawson 1980.

51. Fischer 1984, 177ff.

52. See Alvesson 1987, Fischer 1984.

53. Burawoy 1985, 49.

54. Stark 1980, 119.

55. See Heydebrand 1989, Clegg 1990, Mintzberg 1979, Block 1990.

56. Heydebrand 1989, 344-48.

57. Bauman 1988, 225. See also Clegg 1990, Cooper and Burrell 1988, Lyotard 1984.

58. Hirschhorn 1984.

59. Hirschhorn 1984, 97-98.

60. Block 1990.

61. Zuboff 1988.

62. Zuboff 1988, 11.

63. Kanter 1983.

64. Kanter 1991, 74-75 and passim.

65. See Noble 1984; Shaiken 1984; Zimbalist 1979; Kraft 1977, 1987; Garson 1988.

66. Garson 1988.

67. For instance, Haug 1973, 1975; Derber 1982.

68. Piore and Sabel 1984.

69. Piore and Sabel 1984, 252.

70. Piore and Sabel 1984, 260-61.

71. Piore and Sabel 1984, 261.

72. Reich 1992.

73. Reich 1992, 263.

74. Lash and Urry 1987.

75. Reich 1992, 174ff.

CHAPTER 2

1. Fischer 1990, 66-67.

2. Merchant 1980, 173.

3. Merchant 1980, 180-81.

4. Saint-Simon, as quoted in Markham 1952, 11.

5. Saint-Simon, in Markham 1952, 72-73.

6. Saint-Simon, in Markham 1952, 74-75.

7. Saint-Simon, in Markham 1952, 8.

8. Saint-Simon, in Markham 1952, xxi.

9. Comte, in Markham 1952, xlviii.

10. See Fischer 1990, 74ff., for a fuller discussion.

11. Taylor 1913, 25.

12. Taylor 1913, 129.

13. Taylor (1913, 10) says, for instance, that "prosperity for the employer cannot exist through a long term of years unless it is accompanied by prosperity for the employee, and vice versa; . . . it is possible to give the workman what he most wants—high wages—and the employer what he wants—a low labor cost—for his manufactures."

14. See Fischer 1990, 80-81; Stark 1980.

15. Fischer 1990, 81; see pp. 81ff. for a fuller discussion.

16. Veblen 1921.

17. Veblen 1921, 52.

18. Veblen 1921, 40-41.

19. Veblen 1921, 57.

20. Veblen 1921, 53-54.

21. Veblen 1921, 68.

22. Veblen 1921, 88.

23. Veblen 1921, 165.

24. See Stabile 1986, 42ff.

25. Hoover, quoted in Armytage 1965, 248.

26. See Stabile 1986.

27. Stabile 1986, 49.

28. See Stabile 1986, 50ff., for a fuller discussion.

29. Akin 1977, 25.

30. Akin 1977, 35.

31. Akin 1977, 52-53.

32. Akin 1977, 65.

33. Akin 1977, 82-83.

34. Akin 1977, 83.

35. Akin 1977, 91ff.

36. Akin 1977, 97; see pp. 97ff. for a fuller description of these groups.

37. Akin 1977, 101.

38. Akin 1977, 101.

39. Loeb 1933.

40. Loeb 1933, 7.

41. Loeb 1933, 76.

42. Loeb 1933, 191-92.

43. Loeb 1933, 178.

44. See Fischer 1990, 114 n. 30.

45. See Akin 1977, 110ff; Elsner 1967; Armytage 1965, 238-42.

46. Akin 1977, 170.

47. Burnham 1941.

48. Burnham 1941, 69.

49. Burnham 1941, 75-76.

50. Burnham 1941, 77-78.

51. Reich 1992, 42.

52. Burnham 1941, 87.

53. Burnham 1941, 97-99.

54. He says, for instance, that "managerial economy will have its own

form of crisis. Managerial crises will . . . be technical and political in character: they will result from breakdowns in bureaucratized administration when faced with, say, the complicated problems of sudden shifts to war or peace or abrupt technological changes; or from mass movements of dissatisfaction and revolt, which, with the state and economy fused, would be automatically at once political and economic in character and effect" (Burnham 1941, 125).

55. Burnham 1941, 160.

56. See Bell 1960, 1973; Galbraith 1967; Touraine 1971.

57. Galbraith 1967, 1-4.

58. Galbraith 1967, 6-7.

59. Galbraith 1967, 71.

60. Galbraith 1967, 370.

61. Bell 1973.

62. Bell 1976, xii-xiii.

63. Bell 1976, xiv.

64. Bell 1973, 34.

65. See, for instance, Touraine 1971, 227ff.

66. Gouldner 1979, 65.

67. Block 1990.

68. Block 1990, 11.

69. See chapter 7 for a fuller discussion of these ideas.

70. See Gorz 1968, 1972; Mallet 1975a, 1975b; Touraine 1971.

71. Gorz 1968, 102-4.

72. For Gorz (1968), a loose emphasis on "qualified" workers is usual; for Mallet (1975b), it is all workers in technologically advanced industries; for Touraine (1971), it is primarily intellectuals centered around the university.

73. Mallet 1970, Bahro 1979.

74. Mallet 1970, 61.

75. Ehrenreich and Ehrenreich 1978, Walker 1978.

76. Ehrenreich and Ehrenreich 1978.

77. Ehrenreich and Ehrenreich 1978.

78. Gouldner 1979.

79. See also Chomsky 1982; Gouldner 1975/76, 1979; Putnam 1977.

80. Gouldner 1979.

81. Gouldner 1979, 19. This argument is similar to the Ehrenreich and Ehrenreich analysis of the PMC's appropriation of working-class skill.

82. Gouldner 1979, 24-25.

83. Gouldner 1979, 84-85.

84. See Fischer 1990, 112.

85. Fischer 1990, 112.

86. Foucault 1980, 51-52.

87. Dreyfus and Rabinow 1983, 127.

88. Dreyfus and Rabinow 1983, 134ff.

89. Dreyfus and Rabinow 1983, 195-96.

90. Foucault 1977, 205.

91. See Foucault 1977, 200ff. Also Dreyfus and Rabinow 1983, 188ff.

92. Foucault 1986, 222.

93. Foucault 1980, 96.

94. Foucault 1986, 223.

95. See Foucault 1986, pp. 210ff. on specific rationalities as opposed to the more general concept of rationalization, and pp. 224ff. on strategies of resistance and struggle.

96. Jameson 1991.

97. Lyotard 1984.

98. Lyotard 1984, 4-5.

99. Lyotard 1984, 14.

100. Lyotard 1984, 37.

101. Lyotard 1984, 27.

102. Lyotard 1984, 25-26.

103. Lyotard 1984, 27.

104. Lyotard 1984, 51.

105. Lyotard 1984, 55-56.

106. Lyotard 1984, 67.

107. Lyotard 1984, 60.

CHAPTER 3

1. Shaiken 1984, 7.

2. See Braverman 1974. Neo-Bravermanians include Shaiken 1984, Noble 1984, Zimbalist 1979, Kraft 1977.

3. See Hirschhorn 1984, Susman and Chase 1986, Taylor 1987.

4. See Marglin 1974 for a good discussion.

5. See Braverman 1974.

6. See Feldberg and Glenn 1987; Kraft 1977, 1979, 1987; Noble 1984; Shaiken 1984; Zimbalist 1979.

7. Kraft (1987, 99) exemplifies this perspective when he says, "The computer is a revolutionary tool that will have largely conventional effects in both the workplace and in the society as a whole. . . . the computer will help reproduce, not transform, traditional relations between manager and managed, between those who design and plan and those who carry out the designs and plans of others."

8. See Blauner 1964.

9. Blauner's definition of alienation is less Marxist than psychologistic, emphasizing psychological characteristics: powerlessness, meaninglessness, social alienation, and self-estrangement.

10. Blauner 1964, 182.

11. See Hirschhorn 1984, Salzman 1987, Susman and Chase 1986, Taylor 1987.

12. Hirschhorn (1984, 118) describes the sociotechnical perspective and its origins in this way:

The influential sociotechnical conception of work design was developed at the Tavistock Institute in London. Researchers led by Eric Trist incorporated concepts from anthropology, psychoanalysis, and systems theory. From anthropology these sociotechnical theorists drew methods for mapping out the network of relationships among the material and symbolic elements of a system in which social structure and technique were interrelated. Psychoanalytic insights offered clues for interpreting the interactions among members of small groups or teams. Systems theory indicated that a group could regulate itself without supervision by using feedback on its performance. Finally, Kurt Lewin's field theory, a branch of systems theory, taught that people are motivated to complete tasks—are pulled on by the structure of a task to seek a sense of closure. The three traditions converged to create the concept of the self-regulating work team, completing whole tasks in a work system shaped but not determined by technological parameters and in a social system shaped by psychosocial group processes.

13. See Emery and Trist 1973, Trist et al. 1963.

14. Hirschhorn (1984, 57) argues that in order to maximize the potential for flexibility, the principle of complete integration must be relaxed:

> If the engineer integrates all the production parts into one system and designs control that anticipate all environmental changes, then he comes close to designing the perfect self-regulating machine, the total system that incorporates all relevant forces and processes. He becomes a systems planner and utopian designer....
>
> At the same time, the principle of flexibility is exerting an opposing force. Flexible machinery creates a machine system potential, a capacity to produce many different parts, or combinations of parts, and to change the volume of production. Moreover, as the company's market changes, the machine system's distinctive competence will change. Engineers will develop new software, new control programs, and new configurations of the hardware at hand to adapt the machine to its setting. The postindustrial machine evolves. Its design is open-ended.

15. Hirschhorn 1984, 58.

16. Hirschhorn 1984, 97.

17. Zuboff 1988, 9-10.

18. Noble 1984; see also Noble 1977.

19. Noble 1984, 39.

20. See Noble 1984, 335-36, 339; Shaiken 1984, 53.

21. Noble 1984, 144.

22. Initially, the instructions were given to the machine using punched tape; later the process was computerized.

23. See Shaiken 1984, 47ff.

24. Noble 1984, 150.

25. One designer of N/C made this very clear, pointing out that the manual programming methods had the "unfavorable feature" of leaving "the determination of operation sequences, speeds, feeds, etc. . . . in the hands of the set-up man and the machine operator," thus "the system does not contribute toward close production control by management" (in Noble 1984, 152).

26. Noble (1984, 191-92) contends that this choice was based on politically motivated prejudices of technical, managerial, and military communities. See Noble 1984, 152ff., for a fuller discussion of the political context of R/P and other programming-by-doing approaches. These approaches did find some application in robotics, primarily due to technical difficulties in N/C programming of robots. See also Shaiken 1984, 120ff., on the recent lack of interest in analogic programming—a more decentralized method based on worker skill. Japanese firms have shown some interest in these methods.

27. Shaiken (1984) found that workers felt that "I go home and feel like I haven't done anything" (128) or "On my old job, I controlled the machine; now it controls me" (130). These political conflicts have led to the increasing preference of management for unskilled operators of N/C machines.

28. Shaiken 1984, 119-20.

29. Kraft 1977, 46ff.

30. Hirschhorn 1984, 114.

31. See Shaiken 1984, 73ff., on this potential for error and the militaristic preference for more complex and expensive systems. In addition to total system failure and downtime, errors can occur when a decimal point is lost or added, resulting in a part that is ten times too large or too small.

32. In Shaiken 1984, 125.

33. Braverman 1974, 211-12.

34. Cockburn 1985.

35. Cockburn 1985, 46.

36. Cockburn 1985, 61.

37. Cockburn 1985, 74.

38. Noble 1984, 269.

39. Shaiken 1984, 98.

40. Colwell, quoted in Noble 1984, 333.

41. See Shaiken 1984, 190ff., for an example of such a three-way struggle between managers, workers, and systems analysts when Ford attempted to introduce a new computerized technology. The proposed technology change was eventually abandoned, for a combination of technical and political reasons.

42. See Shaiken 1984, 104ff., for a discussion of an arbitration case over the question of managerial or worker jurisdiction over skill, ultimately decided in favor of the union but circumvented by management. In another case (see pp. 113ff.), a struggle ensued over whether workers would be allowed to learn programming. When they were denied formal training, they learned programming on their own and operated the N/C machines more productively, leading to management censure and an ongoing battle with the official programmers.

43. Shaiken 1984, 110ff.

44. Office of Technology Assessment 1984, 185.

45. Zuboff 1982, 145.

46. Zuboff 1988, 61.

47. Zuboff 1988, 59.

48. Zuboff 1988, 63.

49. Zuboff 1988, 63.

50. Zuboff 1988, 72.

51. Zuboff 1988, 74.

52. Zuboff 1988, 74.

53. Zuboff 1988, 252.

54. Zuboff 1988, 251.

55. Zuboff 1988, 251.

56. Zuboff 1988, 279.

57. Zuboff 1988, 264.

58. One operator described both of these effects, indicating that informating and automating strategies can be combined: "With this new technology it is easier for operators to take on a managerial role because we have more data. There is data every few seconds on everything that is going on. Plus, managers don't have to be standing guard over you to find out what is happening. They can come back in ten days, and from the computer, they can see everything that happened. This eliminates the middleman. It is like going right to the farm to buy your eggs" (Zuboff 1988, 265).

59. Blauner 1964.

60. Kalleberg et al. 1987, 48.

61. Cockburn 1983.

62. In the 1890s, the technology was the "iron comp," which differed from the later linotype machines only in minor ways. See Cockburn 1983, 47ff.

63. See Baron 1987 for a similar analysis of the gendered nature of changes in newspaper compositing.

64. Kalleberg et al. 1987, 57.

65. Cockburn 1983, 52.

66. Cockburn 1983, 108.

67. Cockburn 1983, 103.

68. Cockburn 1983, 116.

69. Cockburn 1983, 107.

70. National Research Council 1986, 31-32. In 1976, women were only 2.1% of compositors in Britain (Cockburn 1983, 162).

71. Cockburn 1983, 118.

72. Cockburn 1983, 143.

73. See Cockburn 1983, 119ff. It appears that the men she interviewed would not have held their jobs as long as they did, and would have been replaced by female operators or eliminated completely, had not their union been strong.

74. See Kalleberg et al. 1987, 58-59.

75. Kalleberg et al. 1987, 60.

76. Kalleberg et al. 1987, 60ff.

77. In Shaiken 1984, 257.

78. Cockburn 1983, 119. See Kalleberg et al. 1987, 62ff., on the way in which these changes have undermined unions.

79. Office of Technology Assessment 1984, 72.

80. Office of Technology Assessment 1984, 60.

81. See Indergaard and Cushion 1987, Office of Technology Assessment 1984, Shaiken 1984. Japan has led the way in superautomated factories. Shaiken (1984, 146) describes an advanced machine tool factory: "The Yamazaki Machinery Corporation, the Japanese machine tool builder, cites some impressive figures for its flexible machine shop in Nagoya, Japan. The $18 million system has 18 machine tools, occupies 30,000 square feet of space, has a staff of 12, and can turn out 74 different products in 1200 variations. A comparable manual system, according to the company, would need 68 machines, 215 employees, and 103,000 square feet to do the same job."

82. See Indergaard and Cushion 1987 for a fuller discussion.

83. This more "cellular" system of subassembly work has replaced the assembly line in workplaces such as Olivetti, Fiat, etc. For a fuller discussion, see Hirschhorn 1988.

84. Shaiken 1984, 168.

85. Indergaard and Cushion 1987, 213.

86. Office of Technology Assessment 1984.

87. Shaiken 1984, 171.

88. In 1980, for instance, the cost of buying and operating a robot for eight years was $6 per hour, whereas the total compensation cost for an autoworker was $20 per hour. See Shaiken 1984, 162-63, for a fuller discussion.

89. See Indergaard and Cushion 1987.

90. Shaiken and Herzenberg 1987.

91. See Shaiken and Herzenberg 1987, 46-47, for a fuller discussion of the technology of this plant.

92. Shaiken and Herzenberg 1987, 49.

93. Shaiken and Herzenberg 1987, 10.

94. Shaiken and Herzenberg 1987, 64.

95. Gallie 1978.

96. Gallie 1978, 295.

97. Gallie 1978, 299.

98. Gallie 1978, 298.

99. For instance, Japanese technological superiority in the steel industry is well known, and Thurow (1988, 293ff.) contends that Japan excels in other industries as well, including many potential "sunrise industries," as evidenced by the fact that Japan has over three times as many robots as the United States.

However, the meaning of this alleged technological superiority is more debatable. Shaiken (1984, 156ff.) contends that given the relatively small numbers of robots in either country, that superiority in robotics cannot begin to explain the considerable differences in productivity and economic competitiveness.

100. See Thurow 1988.

101. Friedman 1988.

102. Friedman 1988, 255.

103. Friedman 1988, 257.

104. Friedman 1988, 257.

105. Friedman 1988, 258.

106. Friedman 1988, 263.

107. Hirschhorn 1984.

108. See Hirschhorn 1984. Problems that Hirschhorn found from his survey of the literature included these: workers not having enough technical knowledge and not obtaining sufficient training; a tension between short-term production exigencies and long-term learning; the uncertainty that accompanied the deemphasis on bureaucratic rules and regulations; the difficulties that some teams had in learning self-management techniques; and the confusion some managers felt over the difference between coordination and more traditional managerial control.

109. See also MacKenzie and Wajcman 1985.

110. Zuboff (1988), for instance, argues that automating strategies are increasingly dysfunctional, as workers must be given opportunities to acquire more comprehensive skill and knowledge if they are to be able to effectively work with advanced technological systems. However, the sort of informating strategy she envisions was far from being the norm in the workplaces she stud-

ied; it was endorsed as a future plan by a large international bank, and experimented with in a limited fashion in one of the paper mills.

Another reason why workers may not be given the requisite training and working conditions required to maximize technological and worker potential lies in the limitation of certain "sociotechnical" perspectives. Some analyses of sociotechnical "systems" are technocratic and managerial in nature. Sociotechnical analysts typically do not concern themselves with technological design, but rather accept a given technology and a given definition of organizational (managerial) purpose and attempt to design a sociotechnical "system" in accordance with these constraints. Taylor (1987, 223), for instance, characterizes the sociotechnical analyst's mission in the following way: "Socio-technical systems is a continuous management process involving four steps: a) definition of the purpose mission and philosophy; b) analysis of the technical and social subsystems; c) specification of the design recommendations; and d) implementation of improvements in the work process." With such an orientation, concern with quality of work life and worker autonomy will tend to be subordinated to concerns with efficiency, optimal functioning of the technology, and general system imperatives.

111. See Zuboff 1988, Skinner 1988. Skinner's research in twelve major manufacturing firms leads him to conclude that managers are the major impediment to the full realization of the benefits of advanced technology.

112. See Shaiken and Herzenberg 1987, Gallie 1978.

113. See Zuboff 1982, 1988, for fuller discussions of this point.

114. See Cockburn 1983.

115. See Hirschhorn 1984, 167ff., on the decline of middle management and supervisory positions in many firms.

116. Shaiken 1984, 57.

117. Hodson 1988.

118. Cockburn 1987.

119. Shaiken 1984, 8-9.

120. Cockburn 1987, 5, 8.

CHAPTER 4

1. See Beninger 1986.

2. Ilchman 1969; Gouldner 1976, 1979; Goss 1961; Smigel 1964; Mintzberg 1979; Bennis and Slater 1968; Larson 1972/73, 1977.

3. Freidson 1984, 10.

4. See Goss 1961.

5. Gouldner 1979; see also Mortimer and McConnell 1978.

6. Larson 1977.

7. Mintzberg 1979.

8. Mintzberg 1979, 346. See also Bennis 1966 and Burns and Stalker 1961 on the difference between "mechanical" and "organic" organizations, and the importance of the latter for adaptability to uncertain environments.

9. Mintzberg 1979, 437, 442.

10. Mintzberg 1979, 460-61.

11. Ilchman 1969.

12. Ilchman's rational-productivity bureaucracy therefore has affinities with the ideological prehistory of technocracy. He recognizes that his analysis is in the tradition of ideologies of technocracy but eschews the term as "imprecise" and as full of "obloquy, fears, and inflated aspirations" (Ilchman 1979, 474).

13. As Ilchman (1969, 477) puts it:

> [There are] continued social and economic policy consequences of legal and political equality—achieved and aspired to. These have thrust the state, almost everywhere in the world, into an active continuous role to maintain or improve incomes and to insure health, safety, and education of whole populations. "Expert" skills in large numbers are required for this not only in government but also in interest associations; flexible, goal-defined policies give scope to those with these skills.
>
> Simultaneously, the needs to induce changes in the modes of production or to sustain a highly diverse and interdependent economy necessitate rational productive skills in the public and private sectors alike. They are used to insure the supply of the factors of production at predictable and plannable levels, to preserve, extend, or project the parameters of foreign commerce, and to foster industrial innovation through research and development activities.

See Ilchman, n. 23, on the differences between legal-rationality and rational-productivity. He contends that legal-rationality does not include the need for planning, which has become more fundamental to bureaucracy, and that efficiency is not synonymous with productivity.

14. Kanter 1984.

15. Kanter 1983. The high-tech firms were Hewlett-Packard (manufacturer of computers), Wang (manufacturer of word processing equipment), Polaroid, an anonymous computer manufacturer, Honeywell's defense systems division, and General Electric's division that produces advanced medical technology. The more traditional firms were General Motors, a telephone company, an insurance company, and a raw materials refining company.

16. Kanter 1984, 111.

17. Kanter 1984, 122.

18. For instance, Kanter 1984, 122, describes how one young personnel manager without technical background was assigned to a task force, given opportunities to acquire technical skills, and demonstrated such aptitude that she was given the opportunity to move into a more technical area. Generalists were in demand, particularly as coordinators, and overspecialization often led to career blockage (p. 125).

19. Kanter 1983, 27.

20. Kanter 1983, 256-67.

21. Kanter 1983, 261.

22. Kanter 1984, 118.

23. Kanter 1983, 55.

24. See, for instance, Hodson 1988, Kunda 1992.

25. Kanter 1984, 126.

26. Kanter 1983, 180.

27. Kanter 1983, 204ff.

28. See Kanter 1983, 195ff., for a fuller discussion of the action projects, which addressed such issues as how to reorganize production so as to increase productivity and reduce worker alienation and how to improve on-the-job training for both workers and supervisors.

29. Kanter 1983, 188.

30. Hodson 1988, 274. See also Hodson 1985.

31. Hodson 1988, 261.

32. Hodson 1988, 262-63. The average turnover rate for workers was 27%,

and average tenure was around two years; average tenure for engineers was a year and a half.

33. Hodson 1988, 268.

34. Hodson 1988, 268.

35. Colclough and Tolbert 1992.

36. Colclough and Tolbert 1992, 38.

37. Colclough and Tolbert 1992, 131.

38. See Kanter 1983. She predicted convergence because of similarities of socioeconomic environment and political/regulatory environment. However, she felt that decentralization of authority and an orientation toward innovation and change would probably always be more characteristic of high-tech firms.

39. Noyelle 1987.

40. Noyelle 1987, 61. It appears that earlier policies of promotion from within were also characteristic of other service corporations. Murulo (1987), for instance, discusses the history of Aetna, saying, "one of Aetna Life's long-standing personnel policies was to hire outsiders into only the lowest level jobs; lower and middle management was recruited from the ranks" (p. 47). She also discusses similar gender barriers.

41. Noyelle 1987, 89.

42. Noyelle 1987, 47.

43. Noyelle 1987, 85-86.

44. Baran 1987, 41.

45. Baran 1987, 32ff.

46. Baran 1987, 45.

47. Baran 1987, 43.

48. See also National Research Council 1986.

49. Feldberg and Glenn 1987, 92; see also Feldberg and Glenn 1983.

50. Baran 1987.

51. Zuboff 1988.

52. Zuboff 1988, 369. This emphasis on monopolizing information is sim-

ilar to the situation she found in the pulp and paper mills, where managers and systems engineers designed the technology in order to maintain traditional authority relations. One systems designer spoke of his relationship with a manager as a "secret pact," one that was designed to protect the manager from scrutiny at higher levels. Another upper-level manager spoke of the way in which he and the systems engineer designed the system to retain control over divisional managers' access to data: "It's okay for them to have an on-line capability if we control what they see. All they will have is a terminal designed for the purpose of receiving what we release to them. They will not have a keyboard. If they ask for more, then we tell them we don't have the technical capability to do what you asked for. Our technological solution includes the idea that we will transmit to them, and they will never request data from us. This way, we can control what they look at, and they could not just come in and snoop" (344).

53. Zuboff 1988, 139-40.

54. Zuboff 1988, 155.

55. Zuboff 1988, 321.

56. Zuboff 1988, 179-80.

57. Machung 1984. See also Applebaum 1987, National Research Council 1986, 64.

58. Machung 1984, 128.

59. National Research Council 1986, 150.

60. Murphree 1984.

61. Garson 1988, 51-52. On computer monitoring, see also Gregory 1983, Gutek 1983, Machung 1984, Shaiken 1984.

62. *In These Times* Editorial Board 1988, 5.

63. See, for instance, National Research Council 1986; also Garson 1988.

64. Burris 1983.

65. National Research Council 1986.

66. Shaiken 1984.

67. Gregory 1983, 266.

68. See Baran 1987, National Research Council 1986.

69. Mellow 1987, 246.

70. Murulo 1987, 49.

71. Baran 1987, 52.

72. Garson 1981, 35, quotes a manager who graphically illustrated both this polarization and its gender overtones: "We are moving from the pyramid shape to the Mae West. The employment chart of the future will show those swellings on the top and we'll never completely get rid of those big bulges of clerks on the bottom. What we're trying to do right now is pull in that waistline [middle management]."

73. Feldberg and Glenn 1980, 11. Similarly, Machung (1984, 132) found in 1980 that 95% of keypunch operators, but only 22% of systems analysts, were female.

74. Glenn and Tolbert (1987, 322), for instance, using a large national data set, found that 60% of all white men in computer-related occupations were in the top two job categories; conversely, 77% of women of color were concentrated in low-level computer operator and data-entry occupations.

75. Hacker 1979.

76. Colclough and Tolbert 1992, 21.

77. Colclough and Tolbert 1992, 24.

78. Colclough and Tolbert 1992, 21.

79. Noyelle 1987, 15-16; see also Applebaum 1987.

80. Murphree 1984.

81. Women went from 11% to 24% of managers, from 17% to 38% of professionals, and from 38% to 65% of technicians (Baran 1987, 50).

82. Machung 1984, 133.

83. Hirschhorn 1984.

84. Hirschhorn 1984, 146.

85. See Zuboff 1988, 252, 295.

86. Cockburn 1985, 178.

87. Murphree 1984, 155-56.

88. Josefowitz 1983.

89. Gallese 1985.

90. Wacjman 1991.

91. Cockburn 1991, 157.

92. Hearn 1987; see also Hearn 1989.

93. Cockburn 1991, 158.

94. Kanter 1977; see also Cockburn 1991.

95. Cockburn 1991, 141.

96. Zuboff 1988.

97. Zuboff 1988, 371.

98. See Applebaum 1987.

99. See Baran 1987, 56.

100. Pratt, cited in Applebaum 1987.

101. Gregory 1983, 286.

102. See Walby 1986, 207 ff., for a fuller discussion.

103. See Applebaum 1987, Olson 1987.

104. Olson 1987.

105. Applebaum 1987, 299.

106. Gregory 1983, Zimmerman and Horwitz 1983.

107. Gerson 1987.

108. Olson and Primps 1984.

109. See Baran 1987, National Research Council 1986, Noyelle 1987.

110. Dreyfus and Dreyfus 1986, Kaplan et al. 1985.

111. See Dreyfus and Dreyfus 1986; Zuboff 1982, 1988.

112. Dreyfus and Dreyfus (1986) contend that the traditional model is still the norm in Japan, accounting for their superiority of managerial skill.

113. Dreyfus and Dreyfus 1986, 160.

114. As Storey (1983, 95) puts it: "As formal and technical rationality increase . . . managers become victims of their own devices."

115. Dreyfus and Dreyfus 1986, 161.

116. Zuboff 1982, 145.

117. Zuboff 1988, 326.

118. See Dreyfus and Dreyfus 1986, 187ff., for a further discussion of this analogy.

119. Zuboff 1988, 264.

120. See Hodson 1988, Kanter 1983.

121. See Burris 1983, 1989a.

122. Feldman and Milch 1982, Taylor 1984, Straussman 1978.

123. Feldman and Milch 1982, 125.

124. Feldman and Milch 1982, xliv.

125. Feldman and Milch 1982, 233.

126. Taylor 1984, 327, 329.

127. Straussman 1978, 136.

128. Straussman 1978, 14.

129. Bischak, 1987.

130. Bischak, 1989, personal communication.

CHAPTER 5

1. See Freidson 1986 for a fuller discussion.

2. Freidson 1986, 9.

3. Larson 1990, 25; see also Larson 1984.

4. See Freidson 1986, 21-22; Larson 1977, 5.

5. Wilensky 1964, 149, see also Larson 1977, 1980.

6. Freidson 1973.

7. Freidson (1986) challenges this ideal type of professionalism, contending that self-employment has never been the norm for most professinals, and that professional self-employment has not appreciably declined during this century except among some elite professional groups (physicians, lawyers, architects, dentists) that represent only a small percentage of all professionals.

However, Freidson nonetheless admits that a substantial majority of all professionals are currently employed rather than self-employed.

8. See Haug 1973, 1975, 1977; Rothman 1984; Toren 1975.

9. Haug 1977. See also Haskell 1984a, xiii ff., on the intensifying skepticism regarding professionals, as evidenced by public opinion polls, malpractice claims, etc.

10. Haug 1973, 201.

11. Montagna 1968, 143.

12. See Larson 1977, 1980; Derber 1982.

13. Larson 1980, 163ff.

14. Larson 1977, 233. The proletarianization approach is similar to French new-working-class theory (Mallet 1975a, 1975b; Gorz 1968; Touraine 1971), although the latter focuses more on the political implications of the clash between the exigencies of educated labor and those of capitalism.

15. Larson 1977, 188ff.

16. Larson 1977, 199.

17. Derber 1982.

18. Derber 1982, 21.

19. Derber 1982, 7.

20. Derber 1982, 200.

21. Derber, Schwartz, and Magrass 1990.

22. Derber, Schwartz, and Magrass 1990, 181.

23. Engel and Hall 1973, 75.

24. Murphy 1990.

25. Freidson 1984.

26. Freidson 1984, 8.

27. Freidson 1984, 10.

28. Abbott 1988.

29. Abbott 1988, 126ff.

30. Abbott 1988, 183.

31. Abbott 1988, 184.

32. National Research Council 1986, 52.

33. Ginzberg 1987.

34. McKinlay 1982, 52.

35. See Engel and Hall 1973, Haug 1977.

36. National Research Council 1986, 57.

37. McKinlay 1982, 45.

38. See Haug 1977.

39. Freidson 1984, 1986.

40. See Dreyfus and Dreyfus 1986 for a fuller discussion.

41. When MYCIN prescriptions were evaluated by a panel of physicians, 70% of its recommendations were found to be acceptable; INTERNIST was found to be less accurate than clinicians by a rather small margin, however. See Dreyfus and Dreyfus 1985b, 115ff., for a fuller discussion.

42. McKinlay 1982, 54.

43. Dreyfus and Dreyfus 1986, 200.

44. National Research Council 1986, 159.

45. See Starr 1982, 12.

46. Mizrahi 1986.

47. Mizrahi 1986, 103.

48. Mizrahi 1986, 104.

49. McKinlay 1982.

50. Derber 1982, 182-83.

51. Strauss et al. 1985.

52. See Cockburn 1985, 115ff., on the political implications of this capital-intensive approach to medical care, as opposed to a more environmental approach that would look at the patient in context so as to specify environmental sources of illness.

53. Cockburn 1985 concludes, however, that this gender segregation may be breaking down to some extent within X-ray divisions, as radiology is a specialty that attracts female physicians and a small number of male radiographers are found.

54. Cockburn 1985, 121.

55. Cockburn 1985, 123-24.

56. Cockburn 1985, 120.

57. See National Research Council 1986.

58. National Research Council 1986, 56.

59. Starr 1982, 428ff.

60. Smigel 1964.

61. Smigel 1964, 4.

62. Smigel 1964, 343.

63. Smigel 1964, 342.

64. Haug 1973, Rothman 1984.

65. Haug 1973, 221.

66. Rothman 1984.

67. Rothman 1984, 191.

68. Rothman (1984, 202) cites figures showing that new admissions to the bar more than tripled between 1964 and 1979.

69. Rothman 1984, 202.

70. Spangler and Lehman 1982.

71. Spangler and Lehman 1982, 79.

72. Spangler and Lehman 1982, 79.

73. Spangler and Lehman 1982, 93.

74. Spangler and Lehman 1982, 96.

75. Aaronson 1977a, 4.

76. Aaronson 1977a, 3.

77. Aaronson 1977a, 4-6; Aaronson 1977b, viii-x.

78. Aaronson 1977a.

79. Heydebrand 1979; see also Heydebrand and Seron 1990.

80. Heydebrand 1979, 33.

81. Heydebrand and Seron 1990, 3.

82. Heydebrand 1979, 48.

83. Weber, as cited in Heydebrand 1979, 48.

84. Freidson 1984, 163.

85. Dreyfus and Dreyfus 1986, 199.

86. Aaronson 1977a, 1977b.

87. See Baldridge 1971, Callahan 1962, Cremin 1961, Tyack 1974.

88. Bowles and Gintis 1976, Tyack 1974.

89. Burris and Heydebrand 1981, Jencks and Riesman 1968. As early as 1907, for instance, a university professor said: "There is set up within the university an 'administration' to which I am held closely accountable. They steer the vessel and I am one of the crew. I am not allowed on the bridge except when summoned; and the councils in which I participate uniformly begin at the point at which policy is already determined. I am not part of the 'administration' but am used by the 'administration' in virtue of qualities that I may possess apart from my academic proficiencies. In authority, in dignity, in salary, the 'administration' are over me, and I under them" (Veysey 1965, 389).

90. See Tyack 1974, 143.

91. Lazerson and Grubb 1974, Perkinson 1968, Veysey 1965.

92. In 1960, for instance, the Heald Commission urged: "Education could learn from such dynamic industries as chemicals, electronics, petroleum, and even agriculture, where rapid technological and administrative innovation has enabled productivity to rise dramatically" (Newt Davidson Collective n.d., 57).

93. Peterson 1987, 140.

94. See Peterson 1987 for a fuller discussion.

95. Peterson 1987, 141ff.

96. Peterson (1987, 144ff.) concludes that primary school teachers' autonomy is the most threatened by educational technology.

97. See Dreyfus and Dreyfus 1986.

98. Dreyfus and Dreyfus 1986, 132-33.

99. Dreyfus and Dreyfus 1986, 145.

100. See Burris and Heydebrand 1981, Veysey 1965.

101. Mortimer and McConnel 1978, Baldridge 1971; see also Mintzberg 1979.

102. Mortimer and McConnel 1978, 269.

103. Beverly 1982, Bowles and Gintis 1976, Burris 1983.

104. Beverly 1982, 108.

105. Beverly 1982, 110.

106. See Burris and Heydebrand 1981.

107. See Rourke and Brooks 1971, 183 and passim, for examples of how administrative computer technology has promoted centralization.

108. See, for instance, Newt Davidson Collective (n.d.) for predictions of dramatic reductions in faculties and class sizes of two or three thousand.

109. Burris and Heydebrand 1981, 17.

110. See Burris and Heydebrand 1981 for a fuller discussion of the Yeshiva case.

111. Collins 1979, Larson 1977.

112. Haug 1973.

113. Salzman 1987.

114. Hodson 1988.

115. Shaiken 1984.

116. Shaiken 1984, 220.

117. Shaiken (1984, 223) quotes the British Council for Science and Society as saying: "Could human intelligence have arisen independently of the practical needs it served? . . . Could modern science have developed in a society where craftmanship and manual work were regarded as unbefitting the thinker? . . . if, in industrial society, intellectual and manual work come to be finally and completely divorced, there must be a doubt whether this will not destroy the basis on which science and industrial development have themselves been able to flourish."

118. Shaiken 1984, 226, 227.

119. Kraft 1977, 1987.

120. Kraft 1977, 59.

121. Kraft 1987, 103.

122. Kraft 1987, 103.

123. Strober and Arnold 1987; Cockburn 1985; and Hacker 1989, 1990.

124. Strober and Arnold 1987. Similarly, Cockburn (1985) found only a handful of women in skilled jobs in the five U.K. computer engineering firms she studied.

125. Strober and Arnold 1987, 171.

126. Hacker 1989, 1990.

127. Hacker 1990, 113.

128. Hacker 1990, 116.

129. Hacker 1990, 113.

130. Hacker 1989, 47.

131. Hacker 1990, 115.

132. Hacker 1989, 36.

133. Hacker 1989, 35.

134. Foucault 1980.

135. Kunda 1992.

136. Kunda 1992, 56.

137. Kunda (1992, 221) says: "A three-tier system is evident that reflects not only hierarchy, but cultural inclusion: Wage Class 4, Wage Class 2, and temporary workers. Wage Class 4 comprises full members, citizens in good standing of the corporate community. As such, they are clearly both subjects and agents of normative control and primary participants in Tech's cultural life and the rituals associated with it. Members of the other two groups regard the applicability of the organizational ideology to themselves with some skepticism. Despite the prevalent rhetoric of normative control, there appears to be a clear contradiction between the reality they live with and its ideological representation."

138. Kunda 1992, 15.

139. Kunda 1992, 221.

140. Although Kunda does not discuss the relevance of his findings for gender and race, given the disproportionate numbers of white men in the expert sector and women and minorities in the nonexpert sector, his findings regarding cultural exclusion are relevant to gender and race discrimination.

141. Carter and Carter 1981, Freidson 1986, Larson 1980, Kraft 1987, Shaiken 1984.

142. National Research Council 1986, 57.

143. Although vulnerable to such rationalization, less elite professionals have also resisted it so as to maintain advantageous working conditions. Nurses, for instance, have maintained personalized nursing care.

Professionals in less elite settings (e.g., legal services lawyers, interns and residents in large university hospitals) appear to share certain of these aspects of "proletarianization," particularly heavy case loads, but not the same degree of external control of their day-to-day work.

144. See, for instance, Colclough and Tolbert, 1992.

CHAPTER 6

1. See Kanter 1983, 1984.

2. See Zuboff 1988, Hirschhorn 1984.

3. See Hacker 1979, Feldberg and Glenn 1987, National Research Council 1986.

4. National Research Council 1986, 20.

5. Cockburn 1985.

6. See Hirschhorn 1984 for a fuller discussion of the implications of this fact.

7. Shaiken 1984, Noble 1984, Shaiken and Herzenberg 1987.

8. See Baran 1987.

9. Collins 1979, Hodson 1988, Noyelle 1987, Reynolds 1983.

10. Machung 1984, 134; see also Hodson 1988.

11. Kanter 1984.

12. See Mintzberg 1979, Peters and Waterman 1982, Bennis and Slater 1968, Kanter 1984.

13. See Hodson 1988.

14. See Shaiken 1984, 238; also Dreyfus and Dreyfus 1986.

15. Gregory 1983, Gutek 1983, Machung 1984, National Research Council 1986, Shaiken 1984.

16. Zuboff 1982, 151; see also Zuboff 1988.

17. See Elden 1986, Hirschhorn 1984, Kanter 1984, Susman and Chase 1986.

18. See Alvesson 1991.

19. Kanter 1983.

20. See Taylor 1987.

21. Freidson 1984.

22. Spangler and Lehman 1982.

23. Shaiken 1984, Kraft 1987.

24. Dreyfus and Dreyfus 1986, National Research Council 1986.

25. See Cockburn 1985.

26. Heydebrand 1979, 1985; Ilchman 1969.

27. See Shaiken 1984, Murphree 1984.

28. See, for instance, Office of Technology Assessment 1984, National Research Council 1986 for good reviews of the evidence.

29. National Research Council 1986, Shaiken 1984.

30. Cockburn (1983) uses the term *skill restructuring*; Hodson (1988) coined the term *skill disruption*. Given the fact that changes in requisite skills are generally planned, *skill restructuring* is probably the more accurate term. See Hirschhorn 1984 on the new types of alienation and stress.

31. See Feldberg and Glenn 1987 for a fuller discussion.

32. Salzman 1987, 10ff., for instance, found that computer design engineers felt that CAD "did the easy part" of their jobs, making the work more challenging and highly skilled.

33. Shaiken 1984, 98.

34. Cockburn 1983.

35. Zuboff 1982, 145.

36. Hirschhorn 1984, 70; for another sociotechnical analysis, see Susman and Chase 1982.

37. See Baran 1987, National Research Council 1986.

38. Brunander, quoted in Alvesson 1987, 131.

39. See Kraft 1979 on managers' difficulties in managing computer programmers; also Hodson 1988, who found that managers in high-tech firms were perceived as insufficiently knowledgeable concerning the technical details of high-tech production, leading to grievances among workers.

40. Fischer 1984, 188.

41. Kanter 1983, Bourdieu and Passeron 1979, Burris 1983.

42. Taylor 1987.

43. Straussman 1978, Feldman and Milch 1982, Taylor 1984.

44. Alvesson 1987, 226.

45. See Mizrahi 1986 on medical systems; see Burris and Heydebrand 1981 on the educational system.

46. Heydebrand 1979, 38.

47. Shaiken 1984, Cockburn 1983, Burris 1983, National Research Council 1986.

48. See Cockburn 1983.

49. Zuboff 1982, 148.

50. See Alvesson 1991.

51. When workers have become aware of alternative methods of workplace restructuring, they have been highly politicized and involved. Shaiken (1984) discusses several examples of workers fighting for the right to do their own programming of machine tools, challenging the authority of programmers, and in some instances challenging a proposed technological system.

52. Storey 1983, 147.

53. MacKenzie and Wajcman 1985; Noble 1977, 1984.

54. Blumberg 1980.

55. Reich 1992, 197.

56. See Gordon, Edwards, and Reich 1982. Primary jobs, generally prevalent in core companies, involve relatively high levels of skill, high pay, job security, and mobility prospects. Secondary jobs, common in peripheral companies, are poorly paid, have little job security, and are generally dead-end. Subsequent refinements of this simple model differentiate between "independent" primary jobs and "subordinate" primary jobs, with the former being relatively skilled and the latter requiring little skill. For fuller discussions, see Noyelle 1987, 10ff.; Hodson and Kaufman 1982.

57. See Gordon, Edwards, and Reich 1982, 200ff.

58. See Hodson and Kaufman 1982, Noyelle 1987.

59. Noyelle 1987, 100-101.

60. Wright and Martin 1987.

61. Wright and Martin 1987, 24.

62. National Research Council 1986, 123.

63. Noyelle 1987, 119.

64. Burris 1983, Baron and Bielby 1982.

65. Carter and Carter 1981, 478.

66. Bielby and Baron 1986.

67. Bielby and Baron 1986, 777.

68. Bielby and Baron 1986, 782.

69. See Ferguson 1984, Glennon 1979, Keller 1985, Lowe and Hubbard 1983.

70. Although some employers apparently continue to use physical strength requirements as an excuse for gender segregation without justification; Baron and Bielby (1984, 784-85) write: "Our data contain numerous instances in which employer cited heavy lifting as the reason for excluding women from specific jobs, although detailed job analyses revealed that those jobs required *no strenuous physical exertion*."

71. Cockburn 1987, 5, 8.

72. Hacker 1989; see also Weinberg 1987.

73. Ehrenreich and Piven 1984, Kamerman 1984.

74. See Ehrenreich and Piven 1984, 163ff.; Burris 1989b, 1989c.

75. Fernandez-Kelly 1983.

76. Lim 1983, Tiano 1987.

77. Lim 1983.

78. See MacKenzie and Wacjman 1985, Noble 1984, Cockburn 1985.

79. See Feldman and Milch 1982, Kraft 1977, Taylor 1984.

80. Heydebrand 1983, Rhodes 1985.

81. Rhodes 1985, 290.

82. Putnam 1977.

83. Putnam 1977, 409.

84. Regan 1986.

85. Regan 1986, 631.

86. Ernst 1985, 343.

87. Ernst 1985, 347.

88. Burawoy 1985, 15.

89. Burawoy 1985, 163.

90. Burawoy 1985, 163.

91. Stark 1986.

92. Szelenyi 1987, 571.

93. Baylis 1974.

94. Baylis 1974, 263-64.

95. Baylis 1974, 267.

96. Mallet 1970.

97. Bahro 1979.

98. This is similar to Gouldner's 1979 "culture of critical discourse."

99. Konrad and Szelenyi 1979.

100. Konrad and Szelenyi 1979, 209.

101. For a fuller discussion see Kennedy 1990a.

102. Jones and Krause 1991, 240.

103. Lewin, quoted in Kennedy 1990b, 350.

104. Kennedy 1990b, 347.

105. See Kennedy 1990b, 355-59. Kennedy says, for instance, that "civil societies constructed by class alliances between professionals, workers, and peasants will be more universal than those which are a product of an alliance between professionals and the authorities" (357).

106. See Kennedy 1990a, 41.

107. Gouldner 1979, 83. See also Jay 1984, 531.

CHAPTER 7

1. See chapter 1; see also Heydebrand 1989 for a good review of this literature.

2. Zuboff 1988; Kanter 1983. Although Kanter does mention the possibility that "mechanistic" and "organic" types of organization can "exist side by side" (p. 204), this is mentioned in passing, and the implications are not explored.

3. Piore and Sabel 1984, 260; see also pp. 274ff. They say, for instance, that "flexible specialization is predicated on collaboration. And the frequent changes in the production process put a premium on craft skills. Thus the production worker's intellectual participation in the work process in enhanced— and his or her role revitalized. Moreover, craft production depends on solidarity and communitarianism" (278). Piore and Sable qualify their preference for flexible specialization with the concern that it would weaken unions, what they fail to realize is that the majority of workers would find no place here, that their model of flexible specialization is inherently elitist.

4. Piore and Sabel 1984, 279.

5. Reich 1992.

6. Reich 1992, 311-12.

7. Reich 1992, 163.

8. Reich does make a brief comment concerning the possibility of "job

enrichment" for production workers, making them into virtual symbolic analysts, but his analysis is too far removed from the microsocial level of the labor process, and he is correspondingly limited in his capacity to suggest reforms at this level.

9. Reich 1992, 98ff.

10. Reich 1992, 104.

11. Brahm and Jones 1992.

12. Block 1990.

13. Block 1990, 212.

14. See, for instance, Friedman 1988; see also Clegg 1990 for a good review of this literature.

15. Clegg 1990, 181; see also pp. 184ff. for a discussion of Japanese work innovations. Miyoshi and Harootunian 1989 also present a sustained discussion of Japan and postmodernism.

16. See Clegg 1990, 206 and passim.

17. See Clegg 1990, 227ff., for a good history of Swedish economic democracy.

18. Clegg 1990, 232.

19. Clegg 1990, 12.

20. Jameson 1991.

21. See Jameson 1991, chap. 1; also Mandel 1975.

22. Jameson 1991, xiv.

23. Jameson 1991, 44.

24. Jameson 1991, 6.

25. Lyotard 1984.

26. Flax 1990; see also Bordo 1990.

27. Benhabib 1990.

28. Benhabib 1990, 122.

29. Benhabib 1990, 123.

30. Horkheimer and Adorno 1972; Marcuse 1964.

31. Habermas 1981; see also Jay 1984.

32. Habermas 1968a, 1971.

33. Habermas 1976.

34. Habermas 1970.

35. Colclough and Tolbert (1992, 37) conclude from their review of this literature: "It is a common practice among high-tech firms to segregate production workers from highly skilled technicians, engineers, and professionals. The fear seems to be that the freedom and autonomy of engineers and professionals might erode the discipline and efficiency of production workers. . . . there are few direct or systematic ways for workers to communicate problems or ideas encountered in production to system designers."

36. Fischer 1990, 347ff.

37. Fischer 1990, 366.

38. Fischer 1990, 374.

39. See Zwerdling 1980 on the way in which workplace democratization has sometimes resulted in such a "snowball" effect, with empowerment at work leading to dramatic transformations of individuals' lives in various contexts.

40. Habermas 1968b, 61.

REFERENCES

Aaronson, David E., ed. 1977a. *The New Justice: Alternatives to Conventional Criminal Adjudication*. Washington, D.C.: National Institute of Law Enforcement and Criminal Justice.

———. 1977b. *Alternatives to Conventional Criminal Adjudication: Guidebook for Planners and Practitioners*. Washington, D.C.: National Institute of Law Enforcement and Criminal Justice.

Abbott, Andrew. 1988. *The System of Professions: An Essay on the Division of Expert Labor*. Chicago: University of Chicago Press.

Akin, William E. 1977. *Technocracy and the American Dream*. Berkeley: University of California Press.

Alvesson, Mats. 1987. *Organizational Theory and Technocratic Consciousness*. New York: Walter de Gruyter.

———. 1991. "Corporate Culture and Corporatism at the Company Level: A Case Study." *Economic and Industrial Democracy* 12:347-67.

Applebaum, Eileen. 1987. "Restructuring Work: Temporary, Part-Time, and At-Home Employment." In *Computer Chips and Paper Clips*. See Hartmann 1987.

Armytage, W. G. H. 1965. *The Rise of the Technocrats*. London: Routledge & Kegan Paul.

Bahro, Rudolph. 1979. *The Alternative in Eastern Europe*. New York: Schocken.

Baldridge, J. Victor. 1971. *Power and Conflict within the University*. New York: Wiley.

Baran, Barbara. 1987. "The Technological Transformation of White-collar Work." In *Computer Chips and Paper Clips*. See Hartmann 1987.

Baron, Ava. 1987. "Contested Terrain Revisited: Technology and Gender Definitions of Work in the Printing Industry, 1850-1920." In *Women, Work, and Technology*. See B. Wright et al. 1987.

Baron, James, and William T. Bielby. 1982. "Workers and Machines: Dimensions and Determinants of Technical Relations in the Workplace." *American Sociological Review* 47:175-88.

——. 1984. "Organizational Barriers to Gender Equality: Sex Segregation of Jobs and Opportunities." In *Gender and the Life Course*, edited by Alice Rossi, 233-51. New York: Aldine.

Bauman, Zygmunt. 1988. "Is There a Postmodern Sociology?" *Theory, Culture, and Society* 5:217-37.

Baylis, Thomas A. 1974. *The Technical Intelligentsia and the East German Elite.* Berkeley: University of California Press.

Bell, Daniel. 1960. *The End of Ideology.* Glencoe, Ill.: Free Press.

——. 1973. *The Coming of Post-Industrial Society.* New York: Basic Books.

——. 1976. "Introduction." In *The Coming of Post-Industrial Society.* See Bell 1973.

Bendix, Reinhard. 1956. *Work and Authority in Industry.* Berkeley: University of California Press.

Benhabib, Seyla. 1990. "Epistemologies of Postmodernism: A Rejoinder to Jean-Francois Lyotard." In *Feminism/Postmodernism.* See Nicholson 1990.

Beniger, James R. 1986. *The Control Revolution.* Cambridge: Harvard University Press.

Bennis, Warren G. 1966. *Changing Organizations.* New York: McGraw-Hill.

Bennis, Warren G., and Philip Slater. 1968. *The Temporary Society.* New York: Harper Colophon.

Benson, J. Kenneth. 1973. "The Analysis of Bureaucractic-Professional Conflict: Functional vs. Dialectical Approaches." *Sociological Quarterly* 14:376-94.

Beverly, John. 1982. "Higher Education and Capitalist Crisis." In *Professionals as Workers.* See Derber 1982.

Bielby, William T., and James N. Baron. 1986. "Men and Women at Work: Sex

Segregation and Statistical Discrimination." *American Journal of Sociology* 91 (4): 759-99.

Bischak, Gregory. 1987. "Civilian Nuclear Regulation and Public Safety in the United States." Ph.D. diss., New School for Social Research, New York.

Blauner, Robert. 1964. *Alienation and Freedom*. Chicago: University of Chicago Press.

Block, Fred. 1990. *Postindustrial Possibilities: A Critique of Economic Discourse*. Berkeley: University of California Press.

Blumberg, Paul. 1980. *Inequality in an Age of Decline*. New York: Oxford University Press.

Bordo, Susan. 1990. "Feminism, Postmodernism, and Gender Skepticism." In *Feminism/Postmodernism*. See Nicholson 1990.

Bourdieu, Pierre, and Jean Passeron. 1979. *The Inheritors*. Chicago: University of Chicago Press.

Bowles, Samuel, and Herbert Gintis. 1976. *Schooling in Capitalist America*. New York: Basic Books.

Brahm, Richard, and Marc Jones. 1992. "The New International Administration of Capital—and Its Imminent Demise." Paper presented at the Commodity Chains and Global Capitalism Conference, Duke University, Durham, N.C., April.

Braverman, Harry. 1974. *Labor and Monopoly Capital: The Degradation of Work in the Twentieth Century*. New York: Monthly Review Press.

Burawoy, Michael. 1985. *The Politics of Production: Factory Regimes under Capitalism and Socialism*. London: Verso.

Burnham, James. 1941. *The Managerial Revolution*. New York: John Day.

Burns, Tom, and G. M. Stalker. 1961. *The Management of Innovation*. London: Tavistock.

Burris, Beverly. 1983. *No Room at the Top: Underemployment and Alienation in the Corporation*. New York: Praeger.

———. 1986. "Technocratic Management: Social and Political Implications." In *Managing the Labour Process*, edited by David Knights and Hugh Willmott, 166-85. London: Gower.

———. 1989a. "Technocratic Organization and Control." *Organization Studies* 10 (1): 1-22.

———. 1989b. "Technocracy and Gender in the Workplace." *Social Problems* 36 (2): 165-80.

———. 1989c. "Technocratic Organization and Gender." *Women's Studies International Forum* 12 (4): 447-62.

Burris, Beverly, and Wolf Heydebrand. 1981. "Educational Control in the United States." In *New Directions for Higher Education: Management Science Applications to Academic Administration*, edited by J. Wilson, no. 35, 5-25. San Francisco: Jossey-Bass.

Callahan, Raymond E. 1962. *Education and the Cult of Efficiency*. Chicago: University of Chicago Press.

Carter, Michael, and Susan Carter. 1981. "Women's Recent Progress in the Professions, or Women Get a Ticket to Ride after the Gravy Train Has Left the Station." *Feminist Studies* 7 (3): 477-504.

Chomsky, Noam. 1982. *Towards a New Cold War*. New York: Harper & Row.

Clawson, Dan. 1980. *Bureaucracy and the Labor Process*. New York: Monthly Review Press.

Clegg, Stewart R. 1990. *Modern Organizations*. Newbury Park, Calif.: Sage.

Cockburn, Cynthia. 1983. *Brothers: Male Dominance and Technological Change*. London: Pluto Press.

———. 1985. *Machinery of Dominance*. London: Pluto Press.

———. 1987. "Restructuring Technology, Restructuring Gender." Paper presented at the American Sociological Association Meeting, Chicago, 17-21 August.

———. 1991. *In the Way of Women*. Ithaca, N.Y.: ILR Press.

Colclough, Glenna, and Charles M. Tolbert, III. 1992. *Work in the Fast Lane: Flexibility, Divisions of Labor, and Inequality in High-Tech Industries*. Albany: State University of New York Press.

Collins, Randall. 1979. *The Credential Society*. New York: Academic Press.

Cooper, Robert, and Gibson Burrell. 1988. "Modernism, Postmodernism, and Organizational Analysis: An Introduction." *Organization Studies* 9 (1): 91-112.

Cornfield, Daniel, ed. 1987. *Workers, Managers, and Technological Change*. New York: Plenum Press.

Cornfield, Daniel B., Polly A. Phipps, Diane P. Bates, Deborah K. Carter, Trudie W. Coker, Kathleen E. Kitzmiller, and Peter B. Wood 1987. "Office Automation, Clerical Workers, and Labor Relations in the Insurance Industry." In *Workers, Managers, and Technological Change*. See Cornfield 1987.

Cremin, Lawrence A. 1961. *The Transformation of the School: Progressivism in American Education, 1876-1957*. New York: Vintage Books.

Daniels, Arlene Kaplan. 1973. "How Free Should Professions Be?" In *The Professions and Their Prospects*. See Freidson 1973.

Derber, Charles, ed. 1982. *Professionals as Workers*. Boston: G. K. Hall.

Derber, Charles, William A. Schwartz, and Yale Magrass. 1990. *Power in the Highest Degree: Professionals and the Rise of a New Mandarin Order*. New York: Oxford University Press.

Dreyfus, Hubert L., and Stuart Dreyfus. 1986. *Mind over Machine*. New York: Free Press.

Dreyfus, Hubert L., and Paul Rabinow. 1983. *Michel Foucault: Beyond Structuralism and Hermeneutics*. Chicago: University of Chicago Press.

Edwards, Richard. 1979. *Contested Terrain*. New York: Basic Books.

Ehrenreich, Barbara, and John Ehrenreich. 1978. "The Professional-Managerial Class." In *Between Labor and Capital*. See Walker 1978.

Ehrenreich, Barbara, and Frances Fox Piven. 1984. "Women and the Welfare State." In *Alternatives: Proposals for America and the Democratic Left*, edited by Irving Howe, 41-60. New York: Pantheon.

Elsner, Henry. 1967. *The Technocrats: Prophets of Automation*. Syracuse, N.Y.: Syracuse University Press.

Emery, Frederick, and Eric Trist. 1973. *Towards a Social Ecology*. New York: Plenum Press.

Engel, Gloria, and Richard Hall. 1973. "The Growing Industrialization of the Professions." In *The Professions and Their Prospects*. See Freidson 1973.

Ernst, Dieter. 1985. "Automation and the Worldwide Restructuring of the Electronics Industry: Strategic Implications for Developing Countries." *World Development* 13 (3): 333-52.

Etzioni, Amitai. 1965. "Organizational Control Structure." In *Handbook of Organizations*, edited by J. G. March, 650-77. New York: Rand McNally.

Feldberg, Roslyn, and Evelyn Glenn. 1980. "Technology and Work Degradation: Re-examining the Impacts of Office Automation." Unpublished paper, Boston University.

———. 1983. "Technology and Work Degradation: Effects of Office Automation on Women Clerical Workers." In *Machina ex Dea*, edited by Joan Rothschild, 59-78. New York: Pergamon Press.

———. 1987. "Technology and the Transformation of Clerical Work." In *Technology and the Transformation of White-Collar Work*. See Kraut 1987.

Feldman, Elliot J., and Jerome Milch. 1982. *Technocracy versus Democracy: The Comparative Politics of International Airports*. Boston: Auburn House.

Ferguson, Kathy. 1984. *The Feminist Case against Bureaucracy*. Philadelphia: Temple University Press.

Fernandez-Kelly, Maria Patricia. 1983. "Gender and Industry on Mexico's New Frontier." In *The Technological Woman*. See Zimmerman 1983.

Fischer, Frank. 1984. "Ideology and Organization Theory." In *Critical Studies in Organization and Bureaucracy*, edited by Frank Fischer and Carmen Sirianni, 172-90. Philadelphia, Pa.: Temple University Press.

———. 1990. *Technocracy and the Politics of Expertise*. Newbury Park, Calif.: Sage.

Flax, Jane. 1990. "Postmodernism and Gender Relations in Feminist Theory." In *Feminism/Postmodernism*. See Nicholson 1990.

Foucault, Michel. 1977. *Discipline and Punish*. New York: Vintage.

———. 1980. *Power/Knowledge*. New York: Pantheon Books.

———. 1986. "The Subject and Power." In *Michel Foucault*. See Dreyfus and Rabinow 1986.

Freidson, Eliot, ed. 1973. *The Professions and Their Prospects*. Beverly Hills, Calif.: Sage.

Freidson, Eliot. 1984. "The Changing Nature of Professional Control." *Annual Review of Sociology* 10:1-20.

———. 1986. *Professional Powers*. Chicago: University of Chicago Press.

Friedman, David. 1988. "Beyond the Age of Ford: Features of Flexible-System Production." In *The Transformation of Industrial Organization*. See F. Hearn 1988.

Galbraith, John Kenneth. 1967. *The New Industrial State*. Boston: Houghton Mifflin.

Gallese, Liz Roman. 1985. *Women Like Us*. New York: William Morrow.

Gallie, Duncan. 1978. *In Search of the New Working Class*. New York: Cambridge University Press.

Garson, Barbara. 1981. "The Electronic Sweatshop: Scanning the Office of the Future." *Mother Jones*, July, 51-52.

————. 1988. *The Electronic Sweatshop*. New York: Penguin Books.

Gerson, Judith. 1987. "Home-based Clerical Work and the Sexual Division of Labor." Paper presented at the American Sociological Association meeting, Chicago, August.

Ginzberg, Eli. 1987. "Technology, Women, and Work: Policy Perspectives." In *Computer Chips and Paper Clips*. See Hartmann 1987.

Glenn, Evelyn N., and Charles M. Tolbert, III. 1987. "Technology and Emerging Patterns of Stratification for Women of Color: Race and Gender Segregation in Computer Occupations." In *Women, Work, and Technology*. See B. Wright et al. 1987.

Glennon, Linda. 1979. *Women and Dualism*. New Brunswick, N.J.: Transaction Books.

Gordon, David, Richard Edwards, and Michael Reich. 1982. *Segmented Work, Divided Workers*. New York: Cambridge University Press.

Gorz, Andre. 1968. *Strategy for Labor*. Boston: Beacon Press.

————. 1972. "Technical Intelligence and the Capitalist Division of Labor." *Telos* 1:27-41.

Goss, Mary E. W. 1961. "Influence and Authority among Physicians in an Outpatient Clinic." *American Sociological Review* 26 (1): 39-50.

Gouldner, Alvin W. 1975/76. "Prologue to a Theory of Revolutionary Intellectuals." *Telos* 26:3-36.

————. 1976. *The Dialectic of Ideology and Technology*. New York: Oxford University Press.

————. 1979. *The Future of Intellectuals and the Rise of the New Class*. New York: Seabury Press.

Gregory, Judith. 1983. "The Next Move: Organizing Women in the Office." In *The Technological Woman*. See Zimmerman 1983.

Gutek, Barbara. 1983. "Women's Work in the Office of the Future." In *The Technological Woman*. See Zimmerman 1983.

Habermas, Jurgen. 1968a. "Technology and Science as Ideology." In *Toward a Rational Society*, 81-122. Boston: Beacon Press.

————. 1968b. "Technical Progress and the Social Life-World." In *Toward a Rational Society*, 50-61. Boston: Beacon Press.

————. 1970. "Toward a Theory of Communicative Competence." In *Recent Sociology No. 2: Patterns of Communicative Behavior*, edited by Hans Dreitzel, 114-48. New York: Macmillan.

————. 1971. *Knowledge and Human Interests*. Boston: Beacon Press.

————. 1976. *Legitimation Crisis*. Boston: Beacon Press.

————. 1981. "Modernity versus Postmodernity." *Neue German Critique* 22:3-14.

Hacker, Sally. 1979. "Sex Stratification, Technology, and Organizational Change: A Longitudinal Case Study at AT&T." *Social Problems* 26 (5): 539-57.

————. 1989. *Pleasure, Power, and Technology*. Boston: Unwin & Hyman.

————. 1990. *Doing It the Hard Way*. Edited by D. Smith and S. Turner. Boston: Unwin & Hyman.

Hartmann, Heidi, ed. 1987. *Computer Chips and Paper Clips*, vol. 2. Washington, D.C.: National Academy Press.

Haskell, Thomas. 1984a. "Introduction." In *The Authority of Experts*. See Haskell 1984b.

————, ed. 1984b. *The Authority of Experts*. Bloomington: Indiana University Press.

Haug, Marie. 1973. "Deprofessionalization: An Alternate Hypothesis for the Future." *Sociological Review Monograph* 20:195-211.

————. 1975. "The Deprofessionalization of Everyone?" *Sociological Focus* 8:197-213.

————. 1977. "Computer Technology and the Obsolescence of the Concept of Profession." In *Work and Technology*, edited by Marie Haug and Jacques Dofny, 215-28. Beverly Hills, Calif.: Sage.

Hearn, Frank, ed. 1988. *The Transformation of Industrial Organization*. Belmont, Calif.: Wadsworth.

Hearn, Jeff. 1987. *The Gender of Oppression: Men, Masculinity, and the Critique of Marxism*. Brighton, England: Wheatsheaf Books.

———. 1989. *The Sexuality of Organization*. London: Sage.

Heydebrand, Wolf. 1979. "The Technocratic Administration of Justice." *Research in Law and Society* 2:29-64.

———. 1983. "Technocratic Corporatism: Toward a Theory of Occupational and Organizational Transformation." In *Organizational Theory and Public Policy*, edited by Richard Hall and Robert Quinn, 93-114. Beverly Hills, Calif.: Sage.

———. 1985. "Technarchy and Neo-Corporatism: Toward a Theory of Organizational Change under Advanced Capitalism and Early State Socialism." *Current Perspectives in Social Theory* 6:71-128.

———. 1989. "New Organizational Forms." *Work and Occupations* 16 (3): 323-57.

Heydebrand, Wolf, and Carroll Seron. 1990. *Rationalizing Justice: The Political Economy of Federal District Courts*. Albany: State University of New York Press.

Hirschhorn, Larry. 1984. *Beyond Mechanization: Work and Technology in a Postindustrial Age*. Cambridge: MIT Press.

———. 1988. "Developmental Work." In *The Transformation of Industrial Organization*. See F. Hearn 1988.

Hodson, Randy. 1985. "Working in High-Tech: Research Issues and Opportunities for the Industrial Sociologist." *The Sociological Quarterly* 26 (3): 351-64.

———. 1988. "Good Jobs and Bad Management: How New Problems Evoke Old Solutions in High-Tech Settings." In *Sociological and Economic Approaches to Labor Markets*, edited by Paula England and George Farkas, 247-79. New York: Plenum Press.

Hodson, Randy, and Robert L. Kaufman. 1982. "Economic Dualism: A Critical Review." *American Sociological Review* 47:727-39.

Horkheimer, Max, and Theodor Adorno. 1972. *Dialectic of Enlightenment*. New York: Seabury Press.

Iacono, Suzanne, and Rob Kling. 1987. "Changing Office Technologies and Transformations of Clerical Jobs: A Historical Perspective." In *Technology and the Transformation of White-Collar Work*. See Kraut 1987.

Ilchman, Warren F. 1969. "Productivity, Administrative Reform, and Antipolitics: Dilemmas for Developing States." In *Political and Administrative Development*, edited by Robert Braibanti, 472-526. Durham, N.C.: Duke University Press.

Indergaard, Michael, and Michael Cushion. 1987. "Conflict, Cooperation, and the Global Auto Factory." In *Workers, Managers, and Technological Change*. See Cornfield 1987.

In These Times Editorial Board. 1988. "Brave New Office?" *In These Times*, 27 January-4 February, 5.

Jameson, Frederic. 1991. *Postmodernism: Or the Cultural Logic of Late Capitalism*. Durham, N.C.: Duke University Press.

Jay, Martin. 1984. "Habermas and Modernism." *Praxis International* 4 (1): 1-14.

Jencks, Christopher, and David Riesman. 1968. *The Academic Revolution*. Garden City, N.Y.: Doubleday.

Jones, Anthony, and Elliott Krause. 1991. "Professions, the State, and the Reconstruction of Socialist Societies." In *Professions and the State*, edited by Anthony Jones, 225-59. Philadelphia: Temple University Press.

Josefowitz, Natasha. 1983. "Paths to Power in High Technology Organizations." In *The Technological Woman*. See Zimmerman 1983.

Kalleberg, Arne, Michael Wallace, Karyn A. Loscocco, Kevin T. Leicht, and Hans-Helmut Ehm 1987. "The Eclipse of Craft: The Changing Face of Labor in the Newspaper Industry." In *Workers, Managers, and Technological Change*. See Cornfield 1987.

Kamerman, Sheila B. 1984. "Women, Children, and Poverty: Public Policies and Female-Headed Families in Industrialized Countries." *Signs* 10:249-71.

Kanter, Rosabeth Moss. 1977. *Men and Women of the Corporation*. New York: Harper & Row.

―――. 1983. *The Change Masters*. New York: Simon & Schuster.

―――. 1984. "Variations in Managerial Career Structures in High-Technology Firms: The Impact of Organizational Characteristics on Internal Labor Market Patterns." In *Internal Labor Markets*, edited by Paul Osterman, 109-32. Cambridge: MIT Press.

―――. 1991. "The Future of Bureaucracy and Hierarchy in Organizational Theory: A Report from the Field." In *Social Theory for a Changing Society*, edited by Pierre Bourdieu and James Coleman, 63-86. Boulder, Colo.: Westview Press.

Kaplan, Robert, Michael Lombardo, and Mignon Mazique. 1985. "A Mirror for Managers: Using Simulation to Develop Management Teams." *Journal of Applied Behavioral Science* 21 (3): 241-53.

Keller, Evelyn Fox. 1985. *Reflections on Gender and Science*. New Haven, Conn.: Yale University Press.

Kennedy, Michael. 1990a. "The Intelligentsia in the Constitution of Civil Societies and Post Communist Regimes in Hungary and Poland." *Working Paper #45*. Ann Arbor: University of Michigan Comparative Study of Social Transformations.

————. 1990b. *Professionals, Power, and Solidarity in Poland*. New York: Cambridge University Press.

Konrad, George, and Ivan Szelenyi. 1979. *Intellectuals on the Road to Class Power*. New York: Harcourt Brace Jovanovich.

Kouzmin, Alexander. 1980. "Control in Organizational Analysis: The Lost Politics." In *The International Yearbook of Organization Studies, 1979*, edited by David Dunkerly and Graham Salaman, 56-89. Boston: Routledge & Kegan Paul.

Kraft, Philip. 1977. *Programmers and Managers: The Routinization of Computer Programming in the United States*. New York: Springer-Verlag.

————. 1979. "The Industrialization of Computer Programming." In *Case Studies on the Labor Process*. See Zimbalist 1979.

————. 1987. "Computers and the Automation of Work." In *Technology and the Transformation of White-Collar Work*. See Kraut 1987.

Kraut, Robert, ed. 1987. *Technology and the Transformation of White-Collar Work*. Hillsdale, N.J.: Erlbaum.

Kunda, Gideon. 1992. *Engineering Culture*. Philadelphia: Temple University Press.

Larson, Magali S. 1972/73. "Notes on Technocracy: Some Problems of Theory, Ideology, and Power." *Berkeley Journal Of Sociology* 17:1-34.

————. 1977. *The Rise of Professionalism*. Berkeley: University of California Press.

————. 1980. "Proletarianization and Educational Labor." *Theory and Society* 9 (1): 131-76.

————. 1984. "The Production of Expertise and the Constitution of Expert Power." In *The Authority of Experts*. See Haskell 1984b.

————. 1990. "In the Matter of Experts and Professionals, Or How Impossible It Is to Leave Nothing Unsaid." In *The Formation of Professions: Knowledge, State, and Strategy*, edited by Rolf Torstendahl and Michael Burrage, 24-50. Newbury Park, Calif.: Sage.

Lash, Scott, and John Urry. 1987. *The End of Organized Capitalism*. Madison: University of Wisconsin Press.

Lazerson, Marvin, and W. Norton Grubb. 1974. *American Education and Vocationalism*. New York: Teachers College Press.

Lim, Linda. 1983. "Capitalism, Imperialism, and Patriarchy: The Dilemma of Third-World Women Workers in Multinational Factories." In *Women, Men, and the International Division of Labor*, edited by June Nash and Maria Patricia Fernandez-Kelly, 70-91. Albany: State University of New York Press.

Loeb, Harold. 1933. *Life in a Technocracy*. New York: Viking Press.

Lowe, Marian, and Ruth Hubbard. 1983. *Woman's Nature: Rationalizations of Inequality*. New York: Pergamon Press.

Lyotard, Jean-Francois. 1984. *The Postmodern Condition*. Minneapolis: University of Minnesota Press.

Machung, Anne. 1984. "Word Processing: Forward for Business, Backward for Women." In *My Troubles Are Going to Have Trouble with Me*. See Sacks and Remy 1984.

MacKenzie, Donald, and Judy Wajcman. 1985. *The Social Shaping of Technology*. Philadelphia: Open University Press.

Mallet, Serge. 1970. "Bureaucracy and Technocracy in the Socialist Countries." *Socialist Revolution* 1 (3): 44-75.

———. 1975a. *Essays on the New Working Class*. Edited and translated by Dick Howard and Dean Savage. St. Louis: Telos Press.

———. 1975b. *The New Working Class*. Bristol, England: Spokesman Books.

Mandel, Ernest. 1975. *Late Capitalism*. London: New Left Books.

Marcuse, Herbert. 1964. *One-Dimensional Man*. Boston: Beacon Press.

Marglin, Stephen A. 1974. "What Do Bosses Do? The Origins and Functions of Hierarchy in Capitalist Production." *The Review of Radical Political Economy* 6:33-60.

Markham, F. M. H. 1952. *Henri Comte de Saint Simon (1760-1825): Selected Writings*. Oxford, England: Basil Blackwell.

McKinlay, John B. 1982. "Toward the Proletarianization of Physicians." In *Professionals as Workers*. See Derber 1982.

Mellow, Gail. 1987. "Introduction." In *Women, Work, and Technology*. See B. Wright et al. 1987.

Merchant, Carolyn. 1980. *The Death of Nature*. New York: Harper and Row.

Mintzberg, Henry. 1979. *The Structuring of Organizations*. Englewood Cliffs, N.J.: Prentice-Hall.

Miyoshi, Masao, and H. D. Harootunian. 1989. *Postmodernism and Japan*. Durham, N.C.: Duke University Press.

Mizrahi, Terry. 1986. *Getting Rid of Patients: Contradictions in the Socialization of Physicians*. New Brunswick, N.J.: Rutgers University Press.

Montagna, Paul. 1968. "Professionalization and Bureaucratization in Large Professional Organizations." *American Journal of Sociology* 74:138-45.

Montgomery, David. 1979. *Workers' Control in America*. New York: Cambridge University Press.

Mortimer, Kenneth, and T. R. McConnell. 1978. *Sharing Authority Effectively: Participation, Interaction, and Discretion*. San Francisco: Jossey-Bass.

Murphree, Mary C. 1984. "Brave New Office: The Changing World of the Legal Secretary." In *My Troubles Are Going to Have Trouble with Me*. See Sacks and Remy 1984.

Murphy, Raymond. 1990. "Proletarianization or Bureaucratization: The Fall of the Professional?" In *The Formation of Professions: Knowledge, State, and Strategy*, edited by Rolf Torstendahl and Michael Burrage, 71-96. Newbury Park, Calif.: Sage.

Murulo, Priscilla. 1987. "White Collar Women and the Rationalization of Clerical Work." In *Technology and the Transformation of White-Collar Work*. See Kraut 1987.

National Research Council. 1986. *Computer Chips and Paper Clips*, vol. 1. Washington, D.C.: National Academy Press.

Newt Davidson Collective. N.d. *Crisis at CUNY*. New York: Newt Davidson Collective.

Nicholson, Linda, ed. 1990. *Feminism/Postmodernism*. New York: Routledge.

Noble, David. 1977. *America by Design: Science, Technology, and the Rise of Corporate Capitalism*. New York: Knopf.

————. 1984. *Forces of Production: A Social History of Industrial Automation*. New York: Knopf.

Noyelle, Thierry J. 1987. *Beyond Industrial Dualism*. Boulder, Colo.: Westview Press.

Office of Technology Assessment. 1984. *Computerized Manufacturing Automation: Employment, Education, and the Workplace*. Washington, D.C.: U.S. Government Printing Office.

Olson, Margrethe. 1987. "Telework: Practical Experience and Future Prospects." In *Technology and the Transformation of White-Collar Work*. See Kraut 1987.

Olson, Margrethe, and Sophia Primps. 1984. "Working at Home with Computers: Work and Non-work Issues." *Journal of Social Issues* 40:97-112.

Perkinson, Henry J. 1968. *The Imperfect Panacea: American Faith in Education, 1865-1965*. New York: Random House.

Peters, Thomas J., and Robert Waterman. 1982. *In Search of Excellence*. New York: Warner Books.

Peterson, Kent D. 1987. "Computerized Instruction, Information Systems, and School Teachers: Labor Relations in Education." In *Workers, Managers, and Technological Change*. See Cornfield 1987.

Piore, Michael, and Charles Sabel. 1984. *The Second Industrial Divide*. New York: Basic Books.

Putnam, Robert D. 1977. "Elite Transformation in Advanced Industrial Societies: An Empirical Assessment of the Theory of Technocracy." *Comparative Political Studies* 19 (3): 383-412.

Reckman, Bob. 1979. "Carpentry: The Craft and Trade." In *Case Studies on the Labor Process*. See Zimbalist 1979.

Regan, Priscilla. 1986. "Privacy, Government Information, and Technology." *Public Administration Review* 63:287-307.

Reich, Robert. 1992. *The Work of Nations*. New York: Vintage.

Reynolds, Diane. 1983. "New Jobs in New Technologies." In *The Technological Woman*. See Zimmerman 1983.

Rhodes, R. A. W. 1985. "Corporatism, Pay Negotiations, and Local Government." *Public Administration* 63:287-307.

Rothman, Robert. 1984. "Deprofessionalization: The Case of Law in America." *Work and Occupations* 11 (2): 183-206.

Rourke, Francis, and Glenn Brooks. 1971. "The Managerial Revolution in Higher Education." In *Academic Governance*, edited by J. Victor Baldridge, 169-92. Berkeley, Calif.: McCutchen.

Sabel, Charles. 1982. *Work and Politics*. New York: Cambridge University Press.

Sacks, Karen, and Dorothy Remy, eds. 1984. *My Troubles Are Going to Have Trouble With Me*. New Brunswick, N.J.: Rutgers University Press.

Salzman, Harold. 1987. "Computer Technology and the Automation of Skill." Paper presented at the American Sociological Association meeting, Chicago, 17-21 August.

Scott, W. Richard. 1966. "Professionals in Bureaucracies: Areas of Conflict." In *Professionalization*, edited by Howard Vollmer and Donald Mills, 265-75. Englewood Cliffs, N.J.: Prentice-Hall.

Shaiken, Harley. 1984. *Work Transformed: Automation and Labor in the Computer Age*. New York: Holt, Rinehart & Winston.

Shaiken, Harley, and Stephen Herzenberg. 1987. *Automation and Global Production: Automobile Engine Production in Mexico, the United States, and Canada*. San Diego: Center for U.S.-Mexican Studies.

Skinner, Wickham. 1988. "Wanted: Managers for the Factory of the Future." In *The Transformation of Industrial Organization*. See F. Hearn 1988.

Smigel, E. O. 1964. *The Wall Street Lawyer: Professional Organization Man?* New York: Free Press.

Spangler, Eve, and Peter Lehman. 1982. "Lawyering as Work." In *Professionals as Workers*. See Derber 1982.

Stabile, Donald. 1986. "Veblen and the Political Economy of the Engineer." *American Journal of Economics and Sociology* 45 (1): 41-52.

Staples, William. 1987. "Technology, Control, and the Social Organization of Work at a British Hardware Firm, 1791-1891." *American Journal of Sociology* 93:62-88.

Stark, David. 1980. "Class Struggle and the Labor Process." *Theory and Society* 9 (1): 89-130.

————. 1986. "Rethinking Internal Labor Markets: New Insights from a Comparative Perspective." *American Sociological Review* 51:492-504.

Starr, Paul. 1982. *The Social Transformation of American Medicine*. New York: Basic Books.

Stone, Katherine. 1974. "The Origins of Job Structure in the Steel Industry." *The Review of Radical Political Economy* 6:61-97.

Storey, John. 1983. *Managerial Prerogative and the Question of Control*. Boston: Routledge & Kegan Paul.

Strauss, Anselm, Shizuko Fagerhaugh, Barbara Suczek, and Carolyn Wiener 1985. *Social Organization of Medical Work*. Chicago: University of Chicago Press.

Straussman, Jeffrey D. 1978. *The Limits of Technocratic Politics*. New Brunswick, N.J.: Transaction Books.

Strober, Myra, and Carolyn Arnold. 1987. "Integrated Circuits/Segregated Labor: Women in Computer-related Occupations and High-tech Industries." In *Computer Chips and Paper Clips*. See Hartmann 1987.

Susman, Gerald, and Richard Chase. 1986. "A Sociotechnical Analysis of the Integrated Factory." *Journal of Applied Behavioral Sciences* 22 (3): 257-70.

Szelenyi, Szonja. 1987. "Social Inequality and Party Membership." *American Sociological Review* 52 (5): 559-73.

Taylor, Frederick W. 1913. *The Principles of Scientific Management*. New York: Harper & Brothers.

Taylor, James C. 1987. "Job Design and Quality of Working Life." In *Technology and the Transformation of White-Collar Work*. See Kraut 1987.

Taylor, Serge. 1984. *Making Bureaucracies Think*. Stanford, Calif.: Stanford University Press.

Thompson, E. P. 1963. *The Making of the English Working Class*. New York: Vintage.

Thurow, Lester. 1988. "Building a World-Class Economy." In *The Transformation of Industrial Organization*. See F. Hearn 1988.

Tiano, Susan. 1987. "Maquiladoras in Mexicali: Integration or Exploitation?" In *Women on the U.S.-Mexico Border: Responses to Change*, edited by Vicki Ruiz and Susan Tiano, 77-103. Boston: Allen & Unwin.

Toren, Nina. 1975. "Deprofessionalization and Its Sources: A Preliminary Analysis." *Sociology of Work and Occupations* 2 (4): 323-37.

Touraine, Alain. 1971. *The Post-Industrial Society*. New York: Random House.

Trist, Eric, G. W. Higgin, H. Murray, and A. B. Pollock 1963. *Organizational Choice*. London: Tavistock.

Tyack, David B. 1974. *The One Best System*. Cambridge: Harvard University Press.

Ulrich, Laurel. 1982. *Good Wives: Image and Reality in the Lives of Women in Northern New England, 1650-1750.* New York: Oxford University Press.

Urban, Michael. 1978. "Bureaucracy, Contradiction, and Ideology in Two Societies." *Administration and Society* 10 (1): 49-85.

Veblen, Thorstein. 1921. *The Engineers and the Price System.* New York: Viking.

Veysey, Laurence R. 1965. *The Emergence of the American University.* Chicago: University of Chicago Press.

Wacjman, Judy. 1991. "Patriarchy, Technology, and Conceptions of Skill." *Work and Occupations* 18 (1): 29-45.

Walby, Sylvia. 1986. *Patriarchy at Work.* Minneapolis: University of Minnesota Press.

———. 1990. *Theorizing Patriarchy.* Oxford: Basil Blackwell.

Walker, Pat. 1978. *Between Labor and Capital: The Professional-Managerial Class.* Boston: South End Press.

Weber, Max. 1978. *Economy and Society,* vol. 2. Edited by Guenther Roth and Claus Wittich. Berkeley: University of California Press.

Weinberg, Sandy. 1987. "Expanding Access to Technology: Computer Equity for Women." In *Women, Work, and Technology.* See B. Wright et al. 1987.

Wilensky, Harold. 1964. "The Professionalization of Everyone?" *American Journal of Sociology* 70 (2): 137-58.

Wright, Barbara, Myra Marx Ferree, Gail Mellow, Linda H. Lewis, Maria-Luz Samper, Robert Asher, and Kathleen Claspell, eds. 1987. *Women, Work, and Technology: Transformations.* Ann Arbor: University of Michigan Press.

Wright, Eric Olin, and Bill Martin. 1987. "The Transformation of the American Class Structure, 1960-1980." *American Journal of Sociology* 93 (1): 1-29.

Zimbalist, Andrew. 1979. *Case Studies on the Labor Process.* New York: Monthly Review Press.

Zimmerman, Jan, ed. 1983. *The Technological Woman.* New York: Praeger.

Zimmerman, Jan, and Jaime Horwitz. 1983. "Living Better Vicariously?" In *The Technological Woman.* See Zimmerman 1983.

Zuboff, Shoshana. 1982. "New Worlds of Computer-Mediated Work." *Harvard Business Review* 60 (2): 142-52.

———. 1988. *In the Age of the Smart Machine.* New York: Basic Books.

Zwerdling, Daniel. 1980. *Workplace Democracy.* New York: Harper Colophon.

INDEX

A

Aaronson, David, 128, 130
Abbott, Andrew, 119
Adhocracy, 15, 81-83, 110, 146, 169
Akin, William, 30
Alienation, worker, 53, 54, 75
American Society of Mechanical
Engineers, 29
Arnold, Carolyn, 138
AT&T, 89-91, 156
Atomic Energy Commission, 110
Attorneys. *See Lawyers*
Automation, 53-56, 60, 61-64, 68-72,
83, 90, 92-96, 103, 124, 135, 145,
148, 161, 169, 178. *See also*
Computers
Automobile industry, 69-71, 74, 89,
156

B

Bacon, Francis, 22
Bahro, Rudolph, 41, 163
Baran, Barbara, 92-93
Baron, James, 158
Baylis, Thomas, 162
Bell, Daniel, 37-39, 40, 42

Benhabib, Seyla, 175-76
Beninger, James, 9
Bentham, Jeremy, 46
Berle, Adolf and Gardner Means, 34
Bielby, William, 158
Bischak, Greg, 109-10
Blauner, Robert, 53, 64, 190n.9
Block, Fred, 16, 39-40, 171-72
Blue-Collar occupations, 20, 36, 52-
79, 148, 150, 157
Blumberg, Paul, 155
Brahm, Richard and Marc Jones, 171
Braverman, Harry, 52, 53, 58
Burawoy, Michael, 161, 162
Bureaucracy, 9-11, 16, 19, 23, 80-112;
bureaucratic conflict, 13;
bureaucratic control, 7, 9, 11, 15,
20; bureaucratic hierarchy, 82;
bureaucratic ideologies, 10;
changes in, 80-112
Burnham, James, 33-35, 49, 187-88n.54
Burris, Beverly, 97

C

Capitalism, 5, 18, 33, 35-38, 40-44, 47,
52, 56, 103, 115, 117, 119, 157, 161-
62, 171, 174